AF711100

Doane's TAX GUIDE FOR FARMERS

by John C. O'Byrne

DOANE

Doane Agricultural Service, Inc. / 8900 Manchester Road / St. Louis, Missouri 63144

Copyright © 1977
by
The Allen Smith Company
Indianapolis, Indiana

Doane's Tax Management Guide
Copyright © 1973
by
The Allen Smith Company
Indianapolis, Indiana

Doane's 1971 Tax Management Guide
Copyright © 1970
by
The Allen Smith Company
Indianapolis, Indiana

PREFACE

The Tax Reform Act of 1976 had such a significant impact on the tax law that we have set out the highlights for quick reference and review.

Congress, the Administration and private interests continually press for legislative changes in the tax law. It is important to be aware of the various influences upon tax legislation and administration of the tax laws. Agriculture, primarily due to the manner in which a "farmer" can report income and expenses, has provided a base for "tax shelter" programs. In some cases the "tax shelter" devices have been used to circumvent or "bend" the tax laws. These alleged abuses created a stimulus for the legislation that became the Tax Reform Act of 1976. In general, the real "working" farmer or rancher escapes most of the anti-shelter provisions but he must be aware of when the new law limitations can have an effect. There are also some beneficial provisions that every farmer and rancher should know about. Chapter 1 summarizes the impact of the new law and guides the reader to the text discussion which places the new law in the context of the old and familiar income tax law.

This edition continues the pattern of prior editions in presenting the income tax law from a practical transactional approach. Emphasis is placed on the tax law as it relates to a farm or ranch transaction. Tax management is stressed because a farmer or rancher must consider the tax effects of a transaction while alternative courses of action are still available. Once an agreement is entered into or a transaction executed, the tax effect is usually fixed.

The current edition continues to recognize the family-agribusiness aspect of modern-day farming and ranching by including discussions of the tax aspects of the various forms of farm and ranch business organization, the estate and gift tax considerations in estate planning for owners of farm and ranch businesses and self-employed retirement plans. The increasing importance of Social Security necessitates a separate chapter on Social Security taxes and benefits and specific problems faced by farm landlords.

We are tremendously indebted to James D. Keast and Charles Davenport, who were co-authors of the 1973 edition. Each of them continues to share in this edition because of their prior contributions. Charles Davenport also participated substantially in the preparation of this edition. Without his efforts, we would not have made it.

DOANE AGRICULTURAL SERVICE, INC.

JOHN C. O'BYRNE

February 1, 1977

THE AUTHOR

JOHN C. O'BYRNE, Dean, Northeastern University Law School.

Born in 1920. M.S. in Public Administration, 1943, Syracuse; J.D., 1948, Harvard. Professor of Law, University of Iowa College of Law, 1948-1966; Professor of Law, Northwestern University, 1966-1972; Dean, Northeastern University School of Law, since 1972. Law practice, Iowa City, Iowa, 1950-1966. Visiting Professor of Law, Boston College Law School, 1965-1966. Farm Editor, Journal of Taxation; Consultant, Internal Revenue Service and Agency for Industrial Development; Director, Corporate Counsel Institute, 1966-1969; Director, Agricultural Law Center, 1956-1963. Co-author, Federal Income Taxation of Partners and Partnerships. Author of articles in legal periodicals and frequent lecturer at tax law institutes and programs.

Dean O'Byrne has attained national recognition as an outstanding authority on tax law.

Additional copies of DOANE'S TAX GUIDE FOR FARMERS may be ordered from Doane Agricultural Service, Inc., 8900 Manchester Road, St. Louis, Missouri 63144.

Readers who seek more information about any point, those who need citations to the statutes, regulations, rulings, and court decisions, and those who prepare a number of farm income tax returns are referred to John C. O'Byrne's **FARM INCOME TAX MANUAL** (Fifth Edition), which may be obtained from the Allen Smith Co., 1435 N. Meridian Street, Indianapolis, Indiana 46202.

TABLE OF CONTENTS

CHAPTER 1

TAX REFORM ACT OF 1976

CHAPTER 2

TAX MANAGEMENT AS A FINANCIAL TOOL

CHAPTER 5

TAX MANAGEMENT OF LAND AND DEPRECIABLE PROPERTY

CHAPTER 6

TAX REPORTING AND RETURNS

CHAPTER 7

SOCIAL SECURITY TAXES AND BENEFITS

CHAPTER 8

TAX ASPECTS OF FARM BUSINESS ORGANIZATION

CHAPTER 9

MINIMIZING TAXATION OF THE FARM ESTATE

chapter 1

TAX REFORM ACT OF 1976

Section

§ 1.1 INTRODUCTION

This chapter discusses the major changes brought about by the Tax Reform Act of 1976. The Act is summarized, and a checklist keyed to the body of the text, is provided. The essential features of the 1976 Act thus are both summarized and integrated into the tax management discussion.

The 1976 Act made numerous changes. Most of them were highly technical. But the dominant theme of the 1976 changes which related directly to farmers was the effort to close tax shelters—largely to those who do appear to be nonfarmers because their interests in farming might broadly be characterized as passive investors rather than active farmers. The law does not read that way, but that was the purpose. Also the 1976 Act largely eliminated the major plank of the 1969 Act's farm tax shelter program, the excess deductions account.

There were even more changes to nonfarm tax shelters. In addition, there were numerous other changes not specifically directed at farms or tax shelters. Some of these provisions are fairly broad; others are so narrow that only an eel can slip through them. Many of them are discussed in the summary and the tax management text.

Probably the most important changes were in the estate and gift tax provisions. They are touched on briefly in this chapter and discussed in more detail in Chapter 9. The changes are far reaching and will substantially reduce estate and gift taxes. There are a few places that the net was drawn a little bit tighter, but in the main, the estate and gift tax provisions will reduce the tax on thousands of farm estates. At the same time, the "carry-over basis" provisions may revolutionize estate planning.

§ 1.2 CHECKLIST

New Law Provisions Directly Related to Farms:	Summary at:	Discussion at:
Excess deductions account	§ 1.3	§ 5.13
Farm syndicates	§ 1.3	§ 4.1
Accrual accounting for farm partnerships and corporations	§ 1.3	§§ 4.1, 6.1

Crop disaster payments and livestock sold on account of drought	§ 1.4	§§ 3.10, 3.24
"At risk" limitation on farm losses	§ 1.3	§ 4.1
Prepaid feed—Farm syndicates	§ 1.3	§§ 4.1, 4.6
Poultry—Farm syndicates	§ 1.3	§§ 4.1, 5.4
Unemployment insurance	§ 1.5	§§ 4.3, 4.9

General Provisions Applicable to Farms

Limitation on deductions for business use of residence	§ 1.6	§ 4.1
Prepaid interest	§ 1.7	§ 4.5
Limitation on investment interest	§ 1.8	§ 4.5
Holding period for long-term capital gains	§ 1.9	§§ 3.12, 3.13
Limitation on use of capital losses against ordinary income	§ 1.9	§ 3.12
Sales between related controlled corporations	§ 1.10	§ 4.21
Sale of residence by elderly	§ 1.11	§ 3.28
Net operating losses —Carry-back election	§ 1.12	§ 4.18
Partnership rules	§ 1.13	§ 8.3
Basis of property received from decedent	§ 1.14	§§ 5.8, 3.4, 3.27, 8.8
Subchapter S corporations	§ 1.15	§ 8.10
Used property for the investment credit	§ 1.16	§ 5.9
Replacement of condemned real estate	§ 1.17	§§ 3.25, 3.24
Commodity futures	§ 1.18	§ 3.21
Moving expenses	§ 1.19	§ 4.3
Taxes and interest during construction period	§ 1.20	§§ 4.5, 4.9
Conventions outside U.S.	§ 1.21	§ 4.11
Tax incentives for the preservation of historic structures	§ 1.22	§ 4.19
Guarantors of noncorporate debts	§ 1.23	§ 4.20
WIN tax credit	§ 1.24	§ 4.23
Individual retirement accounts	§ 1.25	§§ 4.24, 9.4
First year depreciation	§ 1.26	§ 5.2
Corporate tax rates	§ 1.27	§ 8.9
Child care credit	§ 1.28	
Credit for the elderly	§ 1.29	
Individual tax cuts	§ 1.30	
Partnership promotion and organization expenses	§ 1.31	§ 8.3
Estate and gift taxes	§ 1.32	§§ 9.1, 9.2, 9.3

§ 1.3 FARM TAX SHELTER PROVISIONS

For several years, farm tax shelters have been under attack. The 1976 Act repudiated the 1969 approach by providing that no additions are to be made to excess deduction accounts in years beginning after 1975. Existing EDA's will remain to convert some Section 1231 gain to ordinary income or to be reduced to zero by farm income, but otherwise the EDA dies a quiet death. See § 5.13.

In place of EDA, three new concepts were created to deal with farm tax shelters. Corporations which have receipts of $1,000,000 or more in any year and which are not Subchapter S corporations or family corporations, i.e., 50% or more owned by one family, must adopt accrual accounting and capitalize preproductive expenses of most crops. A partnership having a corporate partner which would be required to adopt these methods on its own must adopt them. Outside the corporate area, farm tax shelters are reached by special rules applicable to farm syndicates and the so-called "at risk" rules. Farm syndicates are any form of enterprise which is subject to state or federal securities regulation or which allocates 35% of the losses to passive or limited investors who have no farm background. Farm syndicates may not claim deductions for feed, seed, fertilizer, and supplies except as consumed. The prepaid feed deduction is out for them. They may not deduct the cost of poultry. They must capitalize preproductive costs (except taxes and interest) of orchards and vineyards. The "at risk" rules are designed to deny losses in excess of the investment in the farm enterprise plus borrowings of the enterprise on which the taxpayer may be held liable. The purpose is to deny deductions for amounts which the taxpayer will never be called on to pay. See § 4.1.

§ 1.4 LIVESTOCK SOLD ON ACCOUNT OF DROUGHT

Income realized on the sales of livestock not used in the trade or business and sold on account of drought may be deferred one year. The amount of gain which may be deferred under this provision is the gain attributable to sales in excess of the sales which the taxpayer would have made if there had been no drought. See § 3.10. This provision also applies to insurance proceeds and, under the 1976 Act, to government payments for destroyed crops.

§ 1.5 UNEMPLOYMENT INSURANCE

Some farmers and their employees have been brought under the Unemployment Insurance system. A farmer who has ten or more employees for twenty or more weeks during the year must pay the federal unemployment tax. Even if these employment levels are not met, a payroll of $20,000 in any one quarter will require the payment of the unemployment tax. Employers of domestics will be subject to unemployment insurance tax if wages to the domestics exceed $1000 in any quarter. See § 4.3.

§ 1.6 LIMITATION ON DEDUCTIONS FOR BUSINESS USE OF RESIDENCE

Deductions for the business use of a residence are now sharply limited. Generally, if a portion of the residence is not used regularly and exclusively as the principal place of business, no deductions will be allowed. While there are some minor exceptions to these rules, the home office deduction seems to have been nearly eliminated. The new rules may well have a serious impact on farmers who have been using a part of their residence for both business and personal purposes. See §§ 4.1 and 2.2-2.

§ 1.7 PREPAID INTEREST

The deduction for prepaid interest has been all but eliminated by the 1976 Act. Except for "points" paid with respect to a home mortgage in an area where there is a custom of charging points, prepaid interest may be deducted only as an accrual-basis taxpayer would. The deduction will be prorated over the loan. See § 4.5.

§ 1.8 LIMITATION ON INVESTMENT INTEREST

The limitation on investment interest has been tightened even further by the 1976 Act. It may now not exceed the amount of investment income and $10,000. There is a further allowance up to $15,000 for interest paid on loans to acquire interests in what are essentially family businesses. The limitation has no effect on personal interest or interest related to the farm business. Interest disallowed in one year may be carried over and deducted in later years subject to the limitations. See § 4.5.

§ 1.9 CAPITAL GAINS

The holding period for assets to qualify for long-term capital gains or losses has been extended. Before 1977, the asset had to be held more than six months. In 1977, an asset will have to be held more than nine months. In 1978 and later years, the asset will have to be held more than one year. This extension of the holding period also applies to Section 1231 assets and timber, coal, and iron ore which are given capital gain treatment under Section 631.

The 1976 Act liberalized the deduction of capital losses against ordinary income. The prior limitation was $1000. In 1977, capital losses may be used to offset $2000 of ordinary income. In 1978, $3000 of ordinary income may be offset by capital losses. There was no change in the rule that $2 of long-term loss offsets only $1 of ordinary income. See §§ 3.12, 3.13.

§ 1.10 SALES BETWEEN RELATED CONTROLLED CORPORATIONS

Sales of depreciable property to a corporation 80% or more owned by the seller had been denied capital gain treatment for some time. This rule has now been extended to sales between two corporations which are 80% or more owned by the same taxpayer or spouse. See § 4.21.

§ 1.11 SALE OF RESIDENCE BY ELDERLY

Gain on the first $20,000 of the sales price of a principal residence by taxpayers over 65 has been excluded from income since 1964. The 1976 Act raises the $20,000 figure to $35,000. If the sales price is over $35,000, a prorata part of the gain is excluded. See § 3.28(b).

§ 1.12 OPERATING LOSSES

The deduction for net operating losses was changed in two respects by the 1976 Act. The five-year carry-over period was extended to seven years. Also previously, a loss had to be carried back before it could be carried forward. Under the new Act, a taxpayer may elect to forego the net operating loss carry-back and elect to carry the loss forward. This election is available for a net operating loss each year it is incurred. See § 4.18.

§ 1.13 NEW PARTNERSHIP RULES

The rules for the allocation of partnership losses were changed in two important respects. First, before the 1976 Act, the partnership's allocation of income and losses to partners was respected unless the principal purpose was evasion of taxes. Under the new Act, a partnership allocation is to be given effect only if it also has substantial economic effect. If not, income and losses are to be allocated in accordance with the partners' interests in the partnership which are to be determined from all of the facts and circumstances.

Under the second new rule, if the partnership is not subject to the "at risk" rules, a partner may not include in his basis for his partnership interest any of the partnership's liabilities for which he has no personal responsibility. Thus, limited partners will be denied the right to use liabilities of the partnership as part of their basis for the partnership interest. Since losses are not deductible in excess of partnership basis, this rule will limit the amount of deductible losses. This rule does not apply to partnerships engaged in real estate activity or to those activities subject to the "at risk" rules, such as farms. See §§ 8.3, 8.5, and 5.8.

§ 1.14 BASIS OF PROPERTY RECEIVED FROM DECEDENT

The rules for determining the basis of property received from a decedent have been drastically altered by the 1976 Act. Under prior law, property took a basis equal to the fair market value at death. Under the 1976 Act, property will have "a carry-over basis" which means that it will have a basis equal to the decedent's basis at the time of his death. For computing loss, the basis is the carry-over basis. However, for computing gain or depreciation, the carry-over basis is adjusted by two other factors. The increase in the value of the property which occurred **prior** to December 31, 1976 and federal and state death taxes paid on increases in the value of the property which occurred **after** 1976 are added to the decedent's basis. The adjustment for death taxes may not increase the basis of the property above its fair market value at the date of death. There are some exclusions from these rules for $10,000 worth of household or personal effects, and for other property up to $60,000. Rules for income in respect of a decedent remain unchanged except for the way in which the deduction for death taxes is computed. See § 5.8.

§ 1.15 SUBCHAPTER S CORPORATIONS

Several rules were changed for Subchapter S corporations. The number of shareholders in a Subchapter S corporation may now be expanded to fifteen if an election to be a Subchapter S corporation has been in effect for five years. Before the five-year period has run, there may be more than ten shareholders (but not more than fifteen) if the number of shareholders in excess of ten received their shares by inheritance. Also, voting trusts and grantor trusts (i.e., the grantor is treated as the owner and **all** of the trust income is taxed to him) may now be shareholders in a Subchapter S corporation without disqualifying it. A testamentary trust has a sixty-day period in which to rid itself of the Subchapter S stock without disqualifying the corporation. Further, a Subchapter S election will not be terminated by the addition of a new shareholder unless within sixty days of the time that the new shareholder becomes a shareholder, the new shareholder affirmatively refuses to consent to the election to be taxed as a Subchapter S corporation. Regulations are to be promulgated explaining how a new shareholder makes the affirmative refusal to consent to the election. See § 8.10. The method of computing previously taxed income where accelerated depreciation has been used was changed.

§ 1.16 THE INVESTMENT CREDIT

The investment credit was liberalized by the Tax Reduction Act of 1975. Those provisions were to expire on June 30, 1976, but they have been extended through 1980 by the 1976 Act. The limit for used property was expanded to $100,000 and the rate for the credit was raised to 10%. See § 5.9.

§ 1.17 REPLACEMENT OF CONDEMNED REAL ESTATE

The 1976 Act extended the period of time during which condemned real estate held for investment or business purposes may be replaced by like kind real estate held for business or investment purposes. The period had been two years after the end of the first year in which any part of the gain was realized. Under the 1976 Act, the period will be the end of the **third** taxable year after the first taxable year in which any part of the gain is realized. See § 3.24.

§ 1.18 COMMODITY FUTURES

Although the long-term capital gain period was extended for assets generally, commodity futures remain under the old rule and must only be held more than six months. See § 3.21.

§ 1.19 MOVING EXPENSES

The deduction for moving expenses was liberalized. The old law required a move of more than fifty miles while the new one requires only 35 miles. Costs incurred in searching for new residence, meals and lodgings while occupying temporary quarters in the new location for up to thirty days, expenses for selling the old house and purchasing the new one, or settling an old lease and acquiring a new one have been deductible to the extent of $2500. Only $1000 of that amount could be for house-hunting trips and temporary living quarters. The total of such expenses may now be $3000 and $1500 may be for house-hunting trips and temporary living costs. See § 4.3.

§ 1.20 TAXES AND INTEREST DURING CONSTRUCTION PERIOD

Under IRC Section 266, a taxpayer has an option to capitalize taxes, interest, and other carrying charges on unproductive property. If this unproductive property is real estate, and a construction is started on it, the election to capitalize ceases to exist, and under the 1976 Act, during the construction period, these costs must be capitalized. After they are capitalized, the taxpayer has the option of amortizing them or including them in the basis of the property and depreciating them along with the other depreciable costs. While the election to amortize is quite simple, the mechanics of phasing it in and the amounts allowable during the phase-in period are very complex. During 1976, if this election is chosen, 50% of the taxes and interest may be deducted, and the balance is amortized over the following three years. For more details, see §§ 4.5 and 4.9.

§ 1.21 EXPENSES OF CONVENTIONS OUTSIDE THE UNITED STATES

Expenses of business conventions held outside the United States have been deductible. Under the 1976 Act, the expenses of more than two such conventions will not be deductible in any year. There are also some limitations on the amounts which are deductible even if it is one of the two with respect to which expenses are allowable. See § 4.11.

§ 1.22 TAX INCENTIVES FOR HISTORIC STRUCTURES

Under the 1976 Act, a loss on the demolition of a building which constitutes an historic structure will not be deductible unless there is a certification by the Secretary of the Interior that the building has no significant historical value. If the loss is not allowed, the basis of the property and the cost of demolition must be capitalized as part of the land value. Rapid depreciation or five-year amortization may be applied to certain rehabilitation expenditures on historic structures. These provisions expire in June of 1981. See § 4.19.

§ 1.23 BAD DEBTS OF NONCORPORATE GUARANTORS

Under prior law, a deduction for a bad debt was generally treated as short-term capital loss. However, a guarantor of a debt incurred in the trade or business of the debtor could treat any payment made on the guarantee as an ordinary loss. The 1976 Act changed this rule. Guarantors of all debts are limited to short-term capital loss unless the guarantee was made in the taxpayer's course of business. See § 4.20.

§ 1.24 WIN TAX CREDIT

The 1976 Act made a few changes in the WIN tax credit. See § 4.23.

§ 1.25 INDIVIDUAL RETIREMENT ACCOUNTS

Since 1974, if a taxpayer was not covered by either a qualified employee or a self-employed retirement plan, the taxpayer has been entitled to claim a deduction for an amount put in an individual retirement account up to $1500 a year. Under the 1976 Act, the taxpayer may elect not to take the deduction for his own individual retirement account. Instead, he may take a deduction up to $875 each for himself and his spouse for amounts put into individual retirement accounts for each spouse. A joint account is not an account for each spouse, but there can be a single account with two sub-accounts. See §§ 4.24 and 9.4.

§ 1.26 FIRST YEAR DEPRECIATION

The deduction for first year depreciation by partnerships has been limited to $2000. Previously, it was $2000 for each partner. See § 5.2.

§ 1.27 CORPORATE TAX RATES

The Tax Reduction Act of 1975 lowered the tax rates on corporations to 20% on the first $25,000 of income, 22% on the next $25,000 of income, and 48% on all income thereafter. Those rates were set to expire on June 30, 1976, but the 1976 Act continued them through the end of 1977. See § 8.9.

§ 1.28 CREDIT FOR CHILD CARE EXPENSES

Before 1976, there was a deduction for child care expenses up to $400 a month if two or more children were involved. Under the new law, this deduction has been eliminated and a credit which is subtracted directly from the tax has been substituted for the deduction. The credit is 20% of the child care expenses incurred during the year. The credit may not exceed $400 for one child or $800 for two or more children. The credit is also allowable for expenses attributable to disabled spouses or other disabled dependents. Where both spouses are not fully employed, the maximum credit is computed on the lower of the expenses or the income of the lowest earning spouse. In case of a student, there is a deemed amount of income for credit purposes so that it is allowable in cases where one of the spouses is a student.

§ 1.29 CREDIT FOR THE ELDERLY

The retirement income credit has been replaced with a credit for the elderly, i.e., over 65. The credit is 15% of income up to $2500 in the case of a single person or $3750 in the case of a married couple filing jointly. These amounts are, of course, reduced by Social Security or other nontaxable annuities. Much of the law surrounding the old credit remains for those people who are receiving annuities from public retirement systems and are under the age of 65.

§ 1.30 INDIVIDUAL TAX CUTS

The 1975 Act brought a number of innovations which were carried over to the 1976 Act and are fully effective through calendar year 1977. There is a credit of $35 for each taxpayer and dependent. The credit is subtracted from the tax. If the credit so computed is less than 2% of taxable income up to $9000, a credit equal to 2% of taxable income up to $9000 may be claimed. The low income allowance and standard deduction were both increased for 1975, 1976, and 1977. The low income allowance is now $1700 for a single person, and $2100 for a married couple filing jointly. A married couple filing separately are each entitled to one-half of that amount. The percentage standard deduction was raised to 16% but not to exceed $2800 for a married couple filing jointly and $2400 for a single person. A married person filing separately is entitled to only one-half of the amount allowed to the married couple filing jointly. These allowances and the allowance of the per capita credit are scheduled to expire on December 31, 1977.

§ 1.31 PARTNERSHIP ORGANIZATION AND PROMOTION EXPENSES

Under the 1976 Act, expenses of organizing a partnership and promoting the sale of interests in the partnership must be capitalized. Beginning in 1977, the organization (but not the promotion) expenses may be amortized over sixty or more months. If the partnership terminates before full amortization has occurred, a loss of the balance is allowed. See § 8.3.

§ 1.32 ESTATE AND GIFT TAXES

The estate and gift taxes were substantially revised by the 1976 Act. They are discussed in Chapter 9. The main features of the revisions are:

—A unified rate schedule so that gifts after 1976 and the estate are added together in determining the estate tax.

—A liberalized marital deduction for both the estate and gift taxes.

—A provision limiting generation skipping.

—A special valuation technique for real estate used in farming or other small business.

—Extended periods for payment of the estate tax.

—A requirement that gifts in contemplation of death be grossed up to include the amount of the gift tax paid on them.

—An elimination of the specific exemptions in both the estate and gift taxes and the substitution of a unified credit.

—A revision of the rules on joint tenancy holdings by husband and wife so that if the creation of a joint tenancy after 1976 is treated as a taxable gift, only a half of the joint tenancy will be included in the gross estate of the grantor.

§ 1.33 MINIMUM TAX ON TAX PREFERENCES

The Tax Reform Act of 1976 revised the minimum tax and increased the number of "tax preferences" upon which the tax is levied effective for 1976, i.e., taxable years beginning after 1975.

The minimum tax is 15% (up from 10%) on the amount by which the items of tax preference exceed (1) $10,000 or (2) one half of the regular income tax, whichever is greater.

The items of tax preference subject to the minimum tax are these: The 50% long-term capital gains deduction, and

(1) the amount by which itemized deductions (except medical and casualty) exceed 60% of adjusted gross income,

(2) the amount of fast depreciation in excess of straight line depreciation on real estate,

(3) the amount of fast depreciation in excess of straight line depreciation on leased personal property,

(4) the amount of amortization in excess of depreciation of a pollution control facility,

(5) the amount of depletion in excess of the basis of the depletable property,

(6) the amount by which the amortization deduction for on-the-job training and child care facilities exceeds normal depreciation.

There are other preferences that relate to special deductions for railroad cars, stock options, bad debts of banks, and intangible drilling costs of oil and gas wells. The capital gain preference of a corporation is subject to special rules and there is another special provision which reduces the amount of the preference from capital gains on timber for a corporation.

Partnerships and Subchapter S corporations allocate tax preference items to the partners or shareholders.

These new rules will impose the additional tax on many more taxpayers than did the prior law. For example, if a farmer with $20,000 of taxable income realized a long term capital gain of $50,000 from the sale of capital assets, he would be liable for an additional tax of $2250. This is 15% of $15,000 (half of the gain less the $10,000 exemption).

§ 1.34 MAXIMUM TAX ON PERSONAL SERVICE INCOME

The effect of the maximum tax is to limit the tax on income from personal services to 50%. This means that amounts over $38,000 for a single taxpayer and $52,000 for a married couple filing jointly will be taxed at only 50%.

Personal service income (or earned income under prior law) includes net earnings from self-employment, wages, salaries, professional fees, other compensation for personal services, and pensions and annuities. The personal service income is reduced by the adjusted gross income deductions allocable to such income and by any tax preferences subject to the minimum tax.

The 1976 Act affected the amount subject to the maximum tax because it increased the amount of tax preferences subject to the minimum tax. The amount of income qualified for the maximum tax is reduced dollar for dollar by the amount of tax preferences of the current year.

The changes in the minimum tax on tax preferences are effective for 1976 but the maximum tax changes are effective for 1977. Thus, the tax preference reductions for 1976 follow prior law and are less than they will be for 1977.

chapter 2

TAX MANAGEMENT AS A FINANCIAL TOOL

Section

§ 2.1 THE ASTUTE FARM TAX MANAGER

An astute farm manager "thinks taxes" and maintains careful records so that his day-to-day decisions achieve the greatest tax advantage.

Many farmers and ranchers don't realize how much farm sales would have to increase to provide the same net income as an additional dollar of deduction. If a taxpayer is in the 50% tax bracket, $1 of deduction saves fifty cents. Net sales income (after expenses and cost of goods sold) of $1 will produce fifty cents. What would the gross sales be? $2? $4?

Every farm taxpayer knows that timing of sales and keeping track of deductions have a direct impact upon income. But the tax-wise operator understands that the way he runs his breeding herd determines whether he has capital gains; that the way he sells or trades property affects his gain or loss; that depreciation decisions determine when income will be taxed; that the way he rents his farm controls his Social Security qualification and benefits; that his form of business organization affects income tax results; that investment credits must be paid back if he sells assets too soon; that an election to deduct soil and water conservation or land-clearing expenses may result in recaptured ordinary income on a future sale. As in farming, so in tax, the details make the difference.

A tax-conscious farmer selects the proper time to buy items on which he will take fast depreciation; he sells assets which will recapture depreciation or other previously claimed farm deductions as ordinary income in years of excess deductions; he increases current income to offset an operating loss carry forward; he reviews sales of assets used in farming—land, machinery, livestock, etc.—so that he does not waste capital gains by offsetting them against ordinary losses. In short, he "thinks" taxes.

He plans his sales of capital gain livestock with an eye on the recapture of depreciation rules; and he thinks about the ordinary income recapture of soil and water and land-clearing expenses when he plans a sale of land. If he has an "excess deductions account," his planning pattern is vital; even though he will not add to his EDA after 1975, planning will remain important until the EDA is consumed. Thus, he will try to manage or postpone sales of capital assets that will turn his gain to ordinary income. He plans for short term and long term.

Successful tax management depends upon complete records kept regularly and carefully throughout the year. With good records it is easy to make periodic checks of income and expenses to determine the

1978 SUPPLEMENT
DOANE'S TAX GUIDE FOR FARMERS

The Tax Reduction and Simplification Act and the Social Security Amendments Act were passed during 1977, resulting in several minor changes in the tax laws for 1977 and 1978. Those changes that affect farmers and ranchers are outlined in this supplement and are keyed to specific chapters and paragraphs in Doane's Tax Guide for Farmers.

CHAPTER I. TAX REFORM ACT OF 1976

§1.3. Farm Tax Shelter Provisions

Farm corporations which must use the accrual basis may postpone compliance until 1978 if 65% of the stock is owned by two farm families or 50% by three farm families and the rest of the stock is owned by employees or an employee trust.

§1.24. WIN Tax Credit

There is a new, additional "Jobs Credit" for newly hired employees. The credit is 50% of additional employees' wages up to $4,200.

Another new credit is allowed for hiring handicapped workers.

§1.29. Credit for the Elderly

For 1976 only, a taxpayer may claim the retirement income credit instead of the credit for the elderly.

§1.30. Individual Tax Cuts

The $35 tax credit for 1977 and 1978 is allowed for each exemption, including those for age and blindness. The credits are included in the new tax tables.

The low income allowance and the standard deduction have been superseded by a new "zero bracket amount" of $2,200 for single taxpayers and $3,200 for married couples filing a joint return.

§1.34. Maximum Tax on Personal Service Income

For a farmer, a material part of his income comes from capital as well as personal services. His personal service income is an amount of net profits from farming attributable to personal services, but not in excess of 30% of the net profits.

CHAPTER 3. MANAGING INCOME OF THE FARM BUSINESS

§3.11. Products Donated

Charitable transfers of interests in real estate to conservation organizations in perpetuity are deductible until 1981.

CHAPTER 4. FARM BUSINESS EXPENSES

§4.1. Introduction

Farm corporations which must use the accrual basis may postpone compliance until 1978 if 65% of the stock is owned by two farm families or 50% by three farm families and the rest of the stock is owned by employees or an employee trust.

§4.23. Credit against Tax for WIN Labor

There is a new, additional "Jobs Credit" for newly hired employees. The credit is 50% of additional employees' wages up to $4,200.

Another new credit is allowed for hiring handicapped workers.

CHAPTER 6. TAX REPORTING AND RETURNS

§6.4. Returns which Must Be Filed

New levels for filing returns for 1977 are:

Single, under 65	$2,950
Single, 65 or over	$3,700
Married, filing jointly	$4,700
Married, filing jointly, one 65 or over	$5,450
Married, filing jointly, both 65	$6,200

§6.5. Declarations of Estimated Tax

No penalty will be levied on any part of an underpayment caused by a provision of the 1976 Tax Reform Act.

The penalty interest rate will drop to 6% on February 1, 1978.

CHAPTER 7. SOCIAL SECURITY TAXES AND BENEFITS

§7.2. Employers of Agricultural Labor

For 1978, the wage base increased to $17,700 and the tax rate to 6.05%.

§7.3. Self-Employed Farmers

For 1978, the wage base increased to $17,700 and the tax rate to 8.1%.

§7.4(d). Farm Partners

Limited partnerships can no longer be used as a means of acquiring social security coverage. The distributive share of partnership income or loss received by a limited partner is excluded from social security coverage after 1977.

§7.7. Taxes vs. Benefits upon Retirement

In 1978, a beneficiary between 65 and 72 years of age may earn $4,000 without reduction of social security benefits. For beneficiaries under age 65, the earnings limit for 1978 is $3,240.

The monthly retirement earnings test is eliminated. Thus, it is no longer possible to earn more than the annual maximum ($4,000 in 1978) and still draw full retirement benefits in any month that you can show that you didn't earn more than one-twelfth of the annual maximum ($333). Exception is made for those months of the first year in which you receive benefits.

§7.9. Social Security Benefits and Medicare

Social Security benefits have increased above the figures in the table on page 298.

The Medicare payments by a beneficiary, noted on page 300, have increased.

CHAPTER 8. TAX ASPECTS OF FARM BUSINESS ORGANIZATION

§8.9. Farm Corporation—Regular

The 1977 special rates for corporations are extended through 1978.

CHAPTER 7 SOCIAL SECURITY TAXES AND BENEFITS

approximate taxable income to date and to make business decisions which will result in the greatest net income after taxes.

Since depreciation of farm improvements, machinery, equipment and purchased livestock are important business deductions, farm records should include a detailed record of these capital investments.

There is trouble ahead for the farm manager who does not keep careful records of the deductions for soil and water and land-clearing expenses and the capitalized costs of citrus and almond groves, and the exact details of an "excess deductions account"—for these can lead to tax dangers in the near or far future.

§ 2.2 TAX PLANNING IS A DAILY JOB

§ 2.2-1 How to Manage Income

A cash-basis farmer minimizes taxes by managing sales to keep income balanced from year to year. Whether certain items are sold on December 31 or January 2 might make hundreds of dollars difference in his income tax.

Installment sales and deferred payment sales may postpone all or part of the income from the year of sale until a future year. See § 3.26. Other arrangements may serve to prevent a transaction from being "closed" for income tax purposes. Planning the tax year in which gains are realized may make a substantial tax difference.

An escrow arrangement can be used to postpone income. The deed or bill of sale to the purchaser is left with a third person until stated conditions are met. At that time it is delivered to the purchaser, the deal becomes "closed" and the income is reportable.

Contracts in which the parties have not completed part of the bargain may not be "closed transactions" for tax purposes. When the act occurs which completes the obligations of the parties, then the contract becomes executed and tax liability becomes fixed.

Fixing a sale price in a flat amount plus a percentage of earnings may put off the realization of gain until the payments equal the seller's basis.

Option arrangements which provide for forfeiture of payment on failure to exercise the option will put off the gain or loss until the option is exercised.

Income may sometimes be postponed if the seller receives notes which are not regarded as the equivalent of money. A note may be income if the parties intend it to be and if it has value. On the other hand, a note may be mere evidence of amounts due, not payment at all.

Deferred-Sale Contracts

A cash-basis farmer who wants payment next year for his grain, livestock, or other products delivered this year should have a definite written contract which prevents him from getting his money until next year even though he changes his mind and wants it. The Internal Revenue Service has challenged deferred-payment contracts on the ground that if the contract does not state that it is nonassignable, the farmer can convert the contract claim to cash and therefore should be treated as if he received cash. If the contract is a bona fide, arm's-length sale, provides a small down payment, fixes the dates of subsequent payments and declares that it is non-assignable, the right to income does not arise until the contract date of payment. See § 3.29.

Payment Available

Advance payments for crops and livestock are usually income when received. However, if the transaction is not a "sale," advances may be loans and are included in income only when the sale is made and the obligation to repay is terminated.

When funds are credited to or available to a cash-basis taxpayer without restriction, he is deemed to have received them for income tax purposes. Thus, interest on bank deposits, matured interest coupons on bonds, and dividends on stock are income to a taxpayer regardless of whether he actually collects them. Similarly a check is income when received, not when it is cashed. See § 3.29.

§ 2.2-2 How to Manage Deductions

Management skill in handling deductions is the ability to find all of the legitimate ones, to see a business aspect in what might otherwise be personal, and to have the records to prove the claimed deductions.

All of the ordinary and necessary expenses of the farm business and expenses in connection with the maintenance of income producing property are deductible, but personal expenses are nondeductible unless there is a specific statutory provision permitting deduction. In considering the availability of deductions in planning a course of action, every effort must be made to see the business aspect of a deduction that might appear personal upon preliminary observation. For example, the charging off as a business expense of part of the cost of automobile operation, the food costs attributable to a hired hand, and the business portion of the home telephone are proper applications of the fundamental principle that personal expenses are not deductible, but if the business portion can be carved out, that much is deductible as a business expense.

Before 1976, expenses attributable to a portion of the house used for an office were deductible. Such expenses included depreciation, repairs, utilities and general maintenance. Under the Tax Reform Act of 1976, expenses attributable to business use of a home will be allowed only if a part of the house is exclusively used for such purposes on a regular basis and it is the principal place of business. Amounts, such as taxes and interest, which are allowable as itemized deductions, are not affected by the new law. It would seem that an office in the farm home is the principal place of business. See § 4.1.

Capital Expenditures

Expenses are regarded as current costs, but an expenditure which produces a benefit lasting more than a year is not currently deductible. Such capital expenditures may be recovered through depreciation over the life of the asset. There may be considerable difficulty in ascertaining whether a particular expenditure is a currently deductible "repair" or a capital expenditure. See § 4.1. This ambiguity may, if the taxpayer so elects, be resolved under the so-called ADR regulations by electing to use a **repair allowance.** Amounts falling into this category may be deducted to the extent that they do not exceed a percentage, specified in the ADR regulations, of the unadjusted basis of the assets in a guideline class. Excess expenditures must be capitalized.

The election to use the **repair allowance** is made each year. If it is not elected, the problem of distinguishing between capital expenditures and repairs remains. Each expenditure must be analyzed to determine whether it is properly "current" or not. See § 4.1.

Obviously, this dual system can be advantageous to the farmer. In a year in which he has many expenses which are clearly repairs, he should not choose the repair allowance. In other years when the nature of his expenditures is less clear, he may make the election and claim a deduction for expenditures up to the limit of the repair allowance. This requires annual scrutiny and planning.

Capital expenditures for labor or material to construct buildings or overhaul machinery, or for improvements or rebuilding which add materially to the value or prolong the life of an asset are not repairs but are added to depreciable costs. See § 4.4.

The size of the farm operation will have much to do with drawing the line between expense and capital expenditure. A large farmer's supply item might go unchallenged while the same item on a smaller operator's return should be a depreciable item. Sometimes handy rules of thumb can be based upon dollar values; for example, items costing less than $100 or $200 might be treated as supplies for the average farmer in a particular locality. See §§ 4.4, 4.8.

To Lease or Not to Lease

A farmer may lease an asset used in the farm business and deduct the rental payments as a current business expense. If the transaction is truly a lease, a rental deduction is proper. But the Internal Revenue Service thinks that some leases look very much like conditional sales contracts under which the farmer is in fact buying the asset. If he is buying the asset, the transaction is not a lease and the rental is not deductible but he can properly deduct depreciation and interest or carrying charges.

Careful consideration of the methods of recovering the purchase price of an asset may make the lease charges appear high. A lease is actually another method of borrowing and the lease rental includes interest charge, carrying charge, and depreciation on the property as well as the lessor's overhead and profit. See § 4.1.

Interest

After 1975, interest paid in advance generally will not be deductible by a cash-basis taxpayer when paid but must be prorated over the loan. The Service has been after this one since 1968. "Points" on home mortgages are still currently deductible if there is a community practice of charging them.

If a cash-basis taxpayer pays interest in advance through a discount from the face or principal of the loan, the interest is deductible when the loan is paid. If the loan is payable in installments, he deducts a prorata portion on each repayment.

An accrual-basis taxpayer accrues interest as it is "earned" by the lender whether it is paid in advance or in arrears.

If a contract for the purchase of personal property has a "carrying charge" but no stated interest, 6% of the average unpaid balance may be deducted as interest. "Average unpaid balance" means the sum of the unpaid balances outstanding on the first day of each month during the year divided by twelve. See § 4.5.

If an installment contract does not require at least 6% interest, part of the purchase price will be treated as interest for both the buyer and the seller. See § 3.26.

Large amounts of interest incurred on loans to acquire investment assets (not farm business assets) are subject to a deduction limit. In years between 1970 and 1976, the first $25,000 is deductible and the excess is deductible up to the amount of net income from the investment and the amount of long-term capital gain. Any remaining excess can be deducted only to the extent of one half; the rest is disallowed for the current year but can be carried over to future years. After 1975, the limit is lowered to $10,000 plus investment income.

Excess interest may be carried to later years. This limitation does not apply to personal interest.

Advance Purchases

A balancing of deductions can be achieved by the purchase of supplies, feed and other items before or after the close of a tax year, as the situation requires. For example, there is substantial choice on when to paint or repair buildings, to buy small tools, to increase seedings of legumes and grasses, and to incur soil and water conservation or land-clearing or liming costs. A tax-conscious farmer can easily adjust these to years of high income.

Prepayment for large amounts of feed at the end of the year may be suspect. Cattle, hog and poultry feeders sometimes make a large year-end prepayment of feed to get the deduction in the current tax year. The Internal Revenue Service may take the position that prepayments for feed to be used in the next tax year should not be deducted in the year of payment but should be allocated between the amount used in the year purchased and the amount to be fed in the next taxable year.

However, court decisions indicate that the taxpayer may have the deduction if most of the following facts are present:

(1) The purchase is pursuant to a written contract for an ascertainable amount of feed at a specified price;

(2) The feed is set aside or delivered, and the seller treats the transaction as a sale on his records and books of account;

(3) There is a business purpose, such as a discount in price, for the purchase;

(4) There is no right to a refund of any part of the payment;

(5) There is some reasonable relation between the amount of feed and number of animals to be fed;

(6) The payment is made in cash;

(7) The feed purchase is separate from payment for services. See § 4.6.

The current position of the Internal Revenue Service on deduction of prepaid feed and supplies by a cash-basis taxpayer was established in a ruling issued after these cases. It will use a three pronged test: (1) the payment must be more than a mere deposit; (2) there must be a business reason for the advance payment; and (3) the payment must not materially distort income. If the advance payment fails any one of these tests, the deduction may be taken only as the feed is fed. Obviously, there is some vagueness in these tests, especially the last one. Changes in the pattern of buying feed and supplies, particularly if the change is large in relation to total purchases, would be hard to defend. The Service's position has not been tested in court, but these tests rely heavily on prior case law, except for the last.

Farm syndicates will not be allowed deductions for feed and supplies until they are consumed. After 1975, this rule applies to farm syndicates except those existing before 1976 and which had no change of membership during 1976. Then the rule applies after 1976. A farm syndicate is an organization which sold interests subject to state or federal regulation or where 35% or more of the losses go to persons with limited risks.

Postponing Deductions

The reverse technique is to hold off payment to postpone the cash-basis deduction until the following year. In many cases, it may not matter to the recipient whether he is paid promptly or not. Even if interest is paid on the withheld payment, the interest is deductible.

An accrual-basis taxpayer can add deductions with relative ease as obligations accrue. But he cannot acquire an item and postpone deduction merely by failure to pay. However, if he is not billed for the item, it would normally not be treated as accrued. However, if the item is on hand at the end of the year, it probably must be inventoried and thus offset the deduction to the extent of the inventory amount.

Fees Paid for Tax Advice

Fees and other expenses (for example, the cost of this book) incurred in connection with the determination, collection or refund of any tax are deductible. If the matter has a business aspect, it would be deductible as a business expense, otherwise it should be an itemized personal deduction. For example, the part of an estate planning fee which relates to tax advice is properly deductible. It is always advisable to have the accountant or lawyer allocate total fees to tax advice and other matters.

Automobile Expenses

To determine the amount of car expenses deductible on the business schedule, the total of the automobile expenses for the year is multiplied by the percentage of business use of the car. The percentage of business use is probably best determined by mileage.

Auto transportation costs may be figured at fifteen cents per mile for the first 15,000 miles of business travel, and ten cents per mile for business travel in excess of 15,000 miles per year. The mileage allowance is a substitute for all operating and fixed costs of the automobile including depreciation. See §§ 4.10, 5.2.

Travel Expenses

The cost of any business transportation (not commuting from home to work) is deductible. However, to deduct the cost of meals, the taxpayer must be away overnight from his "tax home." "Home" is regarded as the general area where the farm business is located. Transportation (not meals) is deductible when a farmer works during the day at a temporary or minor location away from the "tax home," for example, a trip to the sale barn. But an overnight trip to buy cattle justifies the deduction of meals, lodging, and transportation. See § 4.11.

If a farmer has regular work or businesses in two or more areas, one must be the tax home. This depends upon the time spent on each job, the business activity in each area and the relative financial return from each.

Travel expenses for farm clinics, short courses, and similar learning activities are deductible. Convention expenses are deductible if there is a relationship to the farm business in the sense that it benefits or advances the farmer's business interests. See § 4.11.

Business Bad Debts

A bad-debt deduction may not be claimed for any item which has not previously been treated as income or which has not actually been paid out. For example, the failure to collect an account due to a cash-basis taxpayer does not create a bad-debt deduction. This tends to restrict the bad-debt deduction to an accrual-basis taxpayer who reported the item for income tax purposes and then failed to collect it.

However, a bad-debt deduction by a cash-basis taxpayer is possible. For example, in connection with a business deal, the farmer might give change in cash for a check which is returned unpaid because of insufficient funds, or a business loan cannot be collected, or a check received in one year and reported as income is not honored the next year. See § 4.20.

Products Donated

The double deduction that farmers previously could obtain by giving raised produce to a charity is gone. Under present law, the charitable deduction for property which would produce ordinary income on sale is limited to its basis. For a cash-basis taxpayer who donates wheat which he raised, the deduction is zero. If the gift is property that would produce long-term capital gain, the rules are rather complicated, but generally there is a full deduction for a gift to a public charity and the deduction is reduced by 50% of the appreciation in value on gifts to other charities. See § 3.4 on gifts of crop-share rents.

§ 2.2-3 How to Manage Business Assets

Long-term capital gains and losses arise from transactions in two separate types of property. A sale or other disposition of a capital asset results in a capital gain or loss. Sales and other dispositions of assets used in the business of farming, and involuntary conversions of capital assets are aggregated together. If the aggregate is a gain it will be a capital gain. If the aggregate of these transactions is a loss, it will usually be an ordinary loss.

An asset produces long-term gain (only half taxed) if it is held for more than six months in 1976, more than nine months in 1977 and more than a year in 1978 and later years.

The Tax Reform Act of 1976 changed the holding period rules that determine when an asset results in long-term gain or loss and when an asset goes into the business aggregate of Section 1231. Before 1977 the required holding period was "more than six months." In 1977, the period is "more than nine months" and in 1978 and later years, it will be "more than one year." Livestock held for draft, breeding, dairy or sporting purposes have a special rule. Cattle and horses must be held for 24 months and other livestock for 12 months.

Assets which are not held for sale in the ordinary course of business nor used in the business are capital assets, for example, a residence or a boat or a share of stock. Only half of the gain on a capital asset, held for the required period, is subject to tax; a loss is subject to some restrictions on deductibility. See §§ 3.12, 3.20.

Generally the gains and losses which are aggregated under the special rule of Internal Revenue Code Section 1231 are those which arise from sales, exchanges, and involuntary conversions of livestock held for draft, breeding, dairy or sporting purposes, land and other assets used in the business, gains and losses from involuntary conversions of capital assets, timber sales under royalty contracts, the cutting of timber, and unharvested crops sold with the land. All of these must be held for the required period.

If the result of aggregating all of these transactions is a net gain, all of the transactions are treated as capital transactions and the net gain is a capital gain. If the net result of the aggregation of all of the transactions is a loss, then each asset is treated as producing ordinary gain or loss. As a practical matter, a net loss from the aggregation of these transactions will be fully deductible.

Thus, it is an obvious advantage to try to realize business gains and losses in different years so that the gains will be capital gains and the losses will be deductible in full.

A special rule applies to casualty and theft gains and losses. Before these are put into the aggregate described above, there is a separate aggregation of the gains and losses resulting from casualties or theft. If there is a net gain, that gain goes into the aggregate of gains and losses from business assets. If the gains and losses from casualties and thefts result in a net loss, the loss is fully deductible and is not included in the aggregation above. See §§ 3.12-3.18.

Sales to Related Taxpayers

There is no capital gain treatment on sales of depreciable assets between spouses or between an individual and his controlled corporation or between two controlled corporations; the gain is ordinary income notwithstanding the capital asset or business asset rules. Ordinary income also results from sales between a partner and a partnership in which the partner has an 80% interest or between two partnerships in which the same persons have an 80% interest. Losses on sales between related taxpayers are also disallowed under much stricter rules. See § 4.21.

Planning Gains and Losses

Fifty per cent of the profits are taxable on sales of land, depreciable assets used in the business and capital assets held for the required period, while 100% of the profits are taxable if the gain on the sale is "short term" (held for a lesser time). It is, therefore, very important to hold such assets for the required period. One day may make a big difference in taxes. See §§ 3.12, 3.20.

Losses which are capital losses offset capital gains. Any extra capital losses then may be used against ordinary income to the extent of $1000 in 1976, $2000 in 1977, and $3000 in 1978. If the losses are long-term losses (resulting from sales of assets held more than six months in 1976, more than nine months in 1977, and more than one year in 1978 and later years), only one-half of them may be used against ordinary income. If the losses are from business property, they offset gains from business property, but any balance will reduce ordinary income.

§ 2.2-4 How to Manage Depreciable Property

The major tax aspects of depreciable property concern 20% depreciation, fast depreciation, and recapture of depreciation and investment credit on sale or other disposition.

Depreciation methods and problems are discussed in Chapter 5. The various aspects of the handling of depreciable property can best be appreciated by examining the "big picture"—an over-all view of all of the different rules with which one must deal. The chart at the end of Chapter 5 summarizes the whole complex of depreciable property and shows the interrelationship of the various rules that influence planning.

If property subject to depreciation is sold at a gain, part of the gain may be the recovery of the depreciation deducted. Section 1245 of the Internal Revenue Code treats as ordinary income on the sale of depreciable tangible property (except buildings) the part of the gain which represents the recovery of previously claimed depreciation. Livestock is subject to recapture of depreciation taken in 1970 and thereafter. Section 1250 recaptures some but not all of the depreciation taken on buildings and other real estate. These rules are discussed in detail in Chapter 5.

The Asset Depreciation Range (ADR) regulations were adopted by the Service in 1971. These regulations provide rules for determining useful life and set forth limitations on when salvage value will be adjusted by the Service. Generally, they provide for useful lives far shorter than actual lives. The system is not mandatory and may be elected by the taxpayer.

Recapture Problems

Wise planning must consider not only the techniques of recovery of the cost of the property but also the possible recapture of previously claimed deductions or credits. See the chart on recapture rules at the end of Chapter 5.

Ordinary depreciation must be taken. The choices lie in the selection of the useful life and determination of salvage value that will affect the amount of the annual deduction.

Substantial decisions turn on the use of the 20% extra depreciation and the fast methods of depreciation. The choice used to be quick or slow recovery of cost and a farmer could adjust his depreciation to a high write-off in the first and the early years of ownership or an even spread of deductions over the expected useful life.

The enactment of the recapture provisions on depreciable assets made a substantial change in tax planning. Since recapture can result in converting several years of depreciation to ordinary income in a single year, the possibility of early disposition of an asset is an important factor in determining how much depreciation to claim.

The investment credit may be recaptured. The "unearned" portion of any investment credit must be added back to the tax in the year in which the asset subject to the credit is sold. A carry-forward of unused investment credits is available but it is subject to some limitations. See Chapter 5 for detailed discussion.

With respect to livestock, depreciation taken in 1970 and thereafter will be recaptured. Recovery in the sales price of pre-1970 depreciation is still capital gain.

Real Estate

As noted in detail in Chapter 5, there were substantial changes in the depreciation rules on real estate wrought by the Tax Reform Act of 1969. On new buildings, depreciation is limited to the 150% declining balance method; the faster methods cannot be used. Used buildings (except certain residential rental units) may be depreciated only by the straight-line method.

The recapture rules also changed. Recapture used to apply only to the fast depreciation in excess of straight line and then only if the property was held for less than ten years. For property held less than ten years, the amount of the depreciation recaptured was 100% if held for twenty months and it declined by one percentage point for each month thereafter. Under present law, all depreciation in excess of straight line taken after 1969 will be recaptured. The declining percentage rule still applies to depreciation taken before 1969. A farmer may change his method of depreciation to straight line to avoid building up additional potential recapture.

Other Recapture Provisions

A good tax manager must consider not only these recapture rules for investment credit and depreciation but also those which recapture soil and water conservation and land-clearing expenses on the sale of land and amounts in a farmer's "excess deductions account" on the sale of any farm asset which ordinarily produces capital gains. See §§ 5.12, 5.13.

Salvage Value

Salvage value as an auditor's issue is less important on "recapture" assets since there is not much purpose in raising that issue any more. In addition, the rule that true salvage may be reduced by 10% of the basis of the asset also avoids salvage value issues. Furthermore, under ADR regulations, salvage will not be adjusted unless it is incorrect by more than 10% of the unadjusted basis of the asset.

The first of these two 10% salvage rules does not apply to livestock. Therefore, that aspect of tax management which looks to the probability of squabbles with the government is concerned mainly with livestock since the "true" salvage value must be used which on livestock is difficult to determine. But if ADR is elected it will not be adjusted unless it is off by more than 10% of the cost. Fast depreciation on business livestock taken before 1970 still reaps capital gain rewards.

§ 2.3 INCOME PLANNING FOR THE FUTURE

§ 2.3-1 Postponement of Income

An exchange or trade-in of property for similar property avoids recognition of gain or loss and recapture of depreciation but the basis of the new property is carried over from the old.

A substantial tax choice arises whenever a farmer discards one piece of property and acquires a substitute. If an old machine is traded on a new one of like kind, no income tax is due. The new machine takes the basis of the old machine plus any cash paid. If the machine is sold and another purchased with the money, there will be a gain or loss on the sale. The amount of the gain which represents the recapture of depreciation will be treated as ordinary income; the rest of the gain will usually be a capital gain and only 50% of the gain will be taxed. A loss will usually be an ordinary loss. Whether it is advantageous to sell and pay taxes on a gain in order to have a higher cost basis on a new machine for depreciation purposes will require a little pencil arithmetic in every case.

If a trade is an "early disposition," the investment credit will be recaptured, but it will be allowed on the new property equal to the basis of the property. If there is sale and a purchase, the recaptured credit will be the same amount, but the credit will be allowed on the entire cost of the new property. This will likely be more than the basis on a trade.

Before selling or trading in an automobile, machine, dairy cow, race horse, breeding animal or other asset used in the business, the probable tax results of each method should be carefully considered, particularly in the light of potential recapture. See Chapter 5.

Exchanges of Farmland

Farm property, including real estate, held for business use or investment, may be exchanged for like property without tax on the gain. This includes an exchange of farmland for urban realty. See § 3.27 for detailed discussion.

If any farm property is traded for similar property and no gain is recognized, the new property is considered as a continuation of the old, and the new property takes over the basis of the old property. When there are potential gains in increased values, the exchange is an effective technique to postpone the payment of tax on the gain.

There are some special recapture problems to be noted. See §§ 3.27, 5.9 to 5.13.

Seller and Buyer

An installment sale of personal property for $1000 or more or of real estate permits the seller to spread his gain over the installment period. It used to be to the seller's advantage to obtain a higher price on assets used in the business (profit half-taxable) and give a low interest rate to the buyer rather than a high interest rate on a lower price.

However, because of the provisions for the recapture as ordinary income of depreciation taken on personal property and certain tangible realty except buildings (see § 5.10), the recapture of excess depreciation on buildings (see § 5.11) and the rule that unstated interest will be carved out of a contract or mortgage (see § 3.26), the ground rules of the planning game have changed.

If the sale is a sale of several assets for a single lump-sum price, both the seller and the buyer must allocate the price to the various assets. Normally, this allocation is a matter of some tension between them. For example, a seller may want to allocate the price to land, growing crops, and buildings depreciated on the straight-line method because these items usually will produce long-term capital gains. On the other hand, grain, animals held for sale, and recaptured depreciation on equipment will produce ordinary income. Land may also produce ordinary income if subject to recapture for soil and water conservation or land-clearing expenses, and any asset (other than buildings) which ordinarily produces capital gain will trigger an "excess deduction account" recapture as ordinary income. See §§ 5.12, 5.13. Also a disposition may result in recapture of the investment credit if the asset was not held for the period used in claiming the credit.

The buyer's situation is almost the reverse. Because interest is fully deductible, a buyer making installment payments, would ordinarily seek a lower cost price with higher interest rate in preference to a higher price and lower interest rate. Interest paid is fully deductible each year but the investment in farmland would not be recoverable until the time of sale. Any amount invested in depreciable property will be recovered only over the lifetime of the asset. If the seller is entitled to a credit for the elderly, low-income allowance, and increased exemptions, he may be amenable to nontaxed ordinary income interest in lieu of capital gain on the purchase price.

In any event, the interest rate on the contract must be 6% ; otherwise the rate will be deemed to be 7%.

A buyer acquiring some property which depreciates rapidly together with some which depreciates slowly will want to allocate as high a price as possible to the rapidly depreciable property and a lower price to that which depreciates slowly to recover his investment as rapidly as possible. Similarly, a buyer of land and buildings should try to establish a relatively high cost for fences, structures, buildings and other depreciable property and a moderate price on the land and the dwelling house. The more that can be fairly allocated to rapidly depreciating items, the better is the buyer's tax position. But, the seller will not agree to a high price for depreciable assets or land if that causes recapture of depreciation or land expenditures as ordinary income, or to a low price on the residence if he plans to postpone or exclude gain on the residence. See § 3.28.

The investment credit plays its part in this transaction. The seller must recapture an investment credit if the asset was not held for the full life upon which the credit was claimed. The buyer will acquire property which may be eligible for the credit, for example, machinery, livestock, fences, tile drains, feeding floors, storage facilities etc. See § 5.9. It is vital to the buyer to have a specific dollar allocation to property which is eligible for the credit.

Other negotiable aspects should be considered. The buyer might seek to allocate part of the purchase price to standing forage to obtain current deductions. Maybe the standing forage could be included within crops sold with the land. See §§ 3.16, 4.5, 4.6.

Before 1976, the buyer might have sought to prepay interest and claim a deduction, but prepaid interest is no longer deductible when paid unless it is "points" on a home mortgage. Instead it will be allowed prorata over the loan period. Because of these obvious conflicts, the contract price fixed by the parties will generally be accepted for tax purposes so long as the allocation is reasonable. Generally, the parties will be bound by the contract, but the Internal Revenue Service may question the allocation to see that it was reasonably and fairly made.

Residences

Gain on the sale (or involuntary conversion) of the farm residence will not be recognized for income tax purposes to the extent that the sale price is invested in a new residence. A special exclusion of gain on the first $35,000 ($20,000 before 1977) of the sale price of a residence is allowed to a taxpayer 65 years of age or over. See § 3.28.

However, if a very small part of the cost of the farm was allocated to the residence when purchased, the seller is in a difficult position if he claims that a very large part of the sale price was received for the residence. If a large part of the sales price is actually attributable to the residence, good planning suggests an independent appraisal and specific allocation in the contract.

Leasing Property

The owner of property retains the right to the deduction for depreciation. Consequently, it might prove advantageous to enter into a long-term lease rather than to sell the property.

If property would be subject to substantial recapture, leasing postpones the day of reckoning. Real property subject to recapture could be leased for the remainder of the ten-year limit on recapture of pre-1970 depreciation. See § 5.11.

§ 2.3-2 Sharing of Income Within the Family

One of the standard principles of income tax planning is the division of income into several small piles to be taxed in the hands of low-bracket taxpayers. Since a husband and wife usually file joint returns, income splitting between them has no advantage. Therefore the splitting of income with children is the better idea for families.

Income-producing property given outright to children will result in the income being taxed to them. Many kinds of property can be held outright by a minor, or property may be held for the minor in trust or by a custodian under a "gifts to minors" statute in every state.

Gifts to Children

Often it is desired to spread the income from the farm operation. This, too, can be done by transferring property to children and renting the property from them or giving them animals which they will raise and sell. The farm unit may be reorganized as a partnership with the children receiving a share by gift and reporting the income from that share.

The Treasury Department will scrutinize carefully any transaction which gives the appearance of tax evasion. Such arrangements must be bona fide. They are certain to be disregarded unless it can be shown that substantial contributions of capital were made by the child partner or that a substantial part of the labor or management is performed by the partner not contributing assets. However, a donee partner's capital may be acquired by gift. If such a partner acquires the capital to invest, or the partnership share itself, from another member of the family, the gift or transfer must be full and complete so that the donor gives up all dominion and control.

Similar scrutiny will be given to owners of corporate shares or farm land received by gift. No income-splitting transfers should be made without legal advice.

Wages paid to children for services rendered are deductible. Children under nineteen years of age or students may be dependents even though they earn more than the amount of the $750 dependency exemption in the year. See § 4.3. The low-income allowance also aids income splitting because a child can earn $2450 without tax. Income from property does not qualify for the low-income allowance if the child is eligible to be claimed as a dependent by the parent.

Recapture Rules

Recapture has not been a problem on gifts of property. The potential depreciation recapture simply passes to the donee with the gift. A similar rule applies to recapture of land expenditures; the donee assumes the responsibility for reporting ordinary income on the sale of the land. However, there is a somewhat different rule if the donor has an "excess deductions account." If the potential gain in the property given away in any year is more than 25% of the potential gain from all of the donor's farm recapture property, then a portion of the "excess deductions account" will pass to each donee. Otherwise, the donor retains his "excess deductions account" and no part of it passes to the donees. While after 1975 farm losses will not be added to EDA, an EDA built up before 1976 remains to convert capital gain to ordinary income until it is consumed in some fashion.

Another problem arises if the gift consists of property on which any investment credit was claimed. The gift will be an early disposition if it occurs before the life on which credit was based has passed, and some part or all of the credit will be recaptured.

The planner may find that these rules can work to the family advantage by passing the recapture of ordinary income to low-bracket taxpayers and out of the hands of the higher bracket donor.

§ 2.3-3 Averaging Income Over Time

There are two tax techniques which spread the tax effect of high income or large losses in the current year over several years. The "income averaging" provision taxes current high income as if it had been received in the current year and in the prior four years. The "net operating loss deduction" offsets a loss from farm operations or from the sale of farm assets against the income of the last three years and the next seven years. A taxpayer may now elect to forego the carry-back and use the loss only in the future.

If income in any year is a fifth more than the average income for the prior four years and if the excess is more than $3000, all the excess income will be taxed under the income averaging provision, at the rate that applies to the first one-fifth of the excess income. In effect, the progressive rate schedule is eliminated on the top four-fifths of the excess income. See § 6.8.

A net loss from the operation of a farm during the taxable year is fully deductible from the farmer's income from any other sources unless the loss is greater than the amount that the farmer has "at risk" in the farm. His cash, property, and any loans on which he is personally liable or has pledged nonfarm property is "at risk" in the farm. To the extent that there remains a net loss after other gross income for the year has been set off, the loss can be carried back to the third, second and first years preceding the year of the loss. If it is not used up by the carry-back, it is carried forward to the years after the year of the loss in chronological order for seven years. If a taxpayer chooses, he may elect to take all of the loss forward and not use any of it as a carry-back.

Loss on the sale of a farm is treated as a farm operating loss. Any net losses on the sale of business assets (depreciable property and real estate) are business losses and become part of a net operating loss.

A casualty loss on property, whether or not connected with the farm business, is included in the net operating loss. Thus, even a fire loss on a residence becomes a net operating loss subject to the carry-back and carry-forward. See § 4.18.

§ 2.4 LIVESTOCK TAX PLANNING

§ 2.4-1 Maximizing Capital Gains

Livestock held for draft, breeding, dairy, or sporting purposes and held for twelve or 24 months, and other livestock used in the farm business held for over six months (nine months in 1977; one year in 1978 and later years) are considered business assets entitled to special capital gain and ordinary loss treatment. Sound tax management dictates efforts to qualify maximum numbers and achieve maximum capital gains.

Breeding and dairy cattle herds provide excellent opportunities for tax minimization. However, preplanning should not ignore sows, boars, ewes, rams and other animals that are used in the business and not held for sale.

The definition of "property used in the trade or business" which is entitled to special capital gains treatment includes livestock, regardless of age, held for draft, breeding, dairy, or sporting purposes. Cattle and horses must be owned for 24 months and other livestock for twelve months, e.g. swine and sheep, to qualify for the special treatment as draft, breeding, dairy, or sporting livestock. An animal which is used in the business but not held for draft, breeding, dairy, or sporting purposes need be owned for more than six months before 1977. To produce capital gain in 1977, such animals must be held more than nine months. In 1978 and later years, the holding period will be more than one year.

Held for Breeding Purposes

Determining whether particular sales of livestock qualify for capital gains treatment depends upon proof that the farmer or rancher held the animals for the required purpose. "Held" generally means the animal was used for the purpose or the farmer intended to so use it. There seems to be no argument about draft animals and little about dairy animals. The addition of the word "sporting" to the statutory definition is intended to cover primarily race and show horses and dogs.

The arguments arise primarily on "held for breeding purposes." This issue turns primarily on the owner's intent to place it in the herd, even though not in the herd when sold. The government has gone through several phases of challenge and come up with various rules of thumb and guidelines varying from the age test to the primary purpose test. The age test held that if the animal was not of sufficient age to bear young then she would not be eligible even though the owner intended to place the animal in the herd. The extension of the holding period for cattle to two years means that the animals are near maturity, so there should be fewer arguments.

The particular importance of qualifying an immature animal for capital gains treatment is that the cost of raising the animal has been written off. The proceeds upon sale are then treated as a capital gain and thus only one half of the gain is taxable.

A herd owner would normally find it advantageous to cull thoroughly and turn his herd on a continuous basis to obtain maximum qualification. Tax considerations should not override the economic factors; there should be a balance between economic and tax considerations.

Bulls, boars, brood sows, and other breeding animals, bought by a farmer for temporary use and sold at a loss within a year (or two years, as the case may be) produce a loss which does not have to be offset against capital gains from other business assets. See § 3.14.

Accrual Basis

Since capital gains treatment is applicable to most sales of draft, breeding, dairy, and sporting animals, it is unprofitable for many farmers to use the accrual accounting method.

The cash-basis farmer treats as a current expense the cost of raising the animal to a productive stage. Consequently, when he sells the animal his "profit" is the difference between the sale price and zero (his "cost" basis). The cash-basis farmer obtains capital gains on the entire sale price of the animal rather than on the difference between the cost of production and the sale price. The accrual-basis farmer who has carried the animal in his inventory, treats as "profit" the difference between the sale price and the last inventory value. Thus, only a part of the accrual-basis farmer's gross proceeds from sales of raised draft, breeding, dairy, and sporting animals will be entitled to capital gains.

The effective way to minimize the disadvantages of accrual accounting on sales of draft, breeding, dairy, and sporting animals would be to change to the cash method. However, consent of the Commissioner is required to make the change. See § 6.1.

A good-faith transfer of the farm business to a partnership which elects to use the cash basis may be a satisfactory answer. See § 6.2. Incorporation will not serve the purpose because the capital gain will not, as it does in the case of partnership, pass as capital gain to the stockholder. See § 6.9. If the corporation has one class of stock and not more than ten shareholders (fifteen in some cases), it can elect to have its income taxed to its shareholders and not to the corporation, and the stockholders will report the capital gains. Such a corporation is commonly called a Subchapter S corporation or a "tax-option corporation." See § 8.10.

If other factors make retention of the accrual method advisable, the accrual-basis farmer should keep out of inventory as many as possible of his draft, breeding, dairy, and sporting animals. However, a transfer of draft, breeding, dairy, or sporting animals from inventory to a fixed asset account requires consent of the Commissioner. The government holds that the election to put breeding animals into inventory or keep them out must be made when the farmer or rancher first begins his livestock operation. He may not elect with respect to each animal each year.

See § 6.3 for explanation of livestock inventories.

Depreciation

Draft, breeding, dairy, and sporting livestock are depreciable assets used in the farming business. Livestock is subject to depreciation, including 20% first-year depreciation and the fast methods of depreciation.

A cash-basis farmer claims depreciation on a purchased animal, and on an animal received by gift or inheritance if there is a basis for the animal. Raised animals cannot be depreciated if the cost of raising the animals has been deducted as an expense.

An accrual-basis farmer may include purchased animals in inventory or they can be held out of inventory and depreciated. If he holds them out of the inventory, he must also exclude raised animals unless he uses the unit livestock price inventory. A raised animal would have a basis for depreciation only if it had been inventoried and later removed to the depreciation schedule.

Methods of Depreciation

The normally recognized methods of depreciation can be used for livestock. These are the "straight-line," "declining-balance," and "sum-of-the-years-digits." The discussions of these methods in § 5.5 are applicable to livestock.

Practice varies on when depreciation on livestock held for business purposes should start. Some authorities take the position that the depreciable useful life of an animal cannot begin before maturity. However, an argument can be made that animals are "put in use" when held for the herd.

Use of the fast methods of depreciation is proper only if the farmer is the original user of the livestock. The government questioned whether the original use of a breeding animal began with the buyer if the seller had bred it once to test it. It seems clear that the present position is that the animal is not "used" in the hands of the buyer if the prior breeding has been limited to that which is required to prove suitability for breeding purposes. If the animal is immature when purchased, the animal should be considered as purchased prior to reaching the commercial breeding stage, hence, "use" would be deemed to have begun with the purchaser.

The fast methods of depreciation (double declining-balance and sum-of-digits) apply only to tangible property with a useful life of three years or more. The useful life of the asset is related to the particular taxpayer rather than to the useful life of the asset as such. For example, if livestock is usually held for only two to three years, the fast depreciation methods would not apply.

The advisability of using fast depreciation depends upon the particular situation. The double declining-balance method offers the fastest return of investment and postpones salvage value problems. Unless the taxpayer chooses to use the Asset Depreciation Range System (ADR), salvage must be determined and deducted before computing depreciation under the sum-of-the-digits and straight-line methods.

The 150% declining-balance method may be applied to used assets. It offers a recovery faster than straight-line on animals that are not new in use to the farmer. See § 5.5.

Livestock are subject to depreciation recapture which will have an influence on the decision to use fast depreciation since it may come back as ordinary income on sale. All depreciation claimed after 1969 is subject to recapture even though the animal was acquired before 1970.

Extra 20% Depreciation

The extra 20% first-year depreciation allowance may be taken on purchased livestock with a useful life of six years. This applies to new and used property so the "used livestock" problem does not arise as long as the six-year test is met.

There has been some question about immature animals and whether the extra allowance is lost if a young heifer is purchased for the breeding herd. The extra allowance should be taken in the year that regular depreciation properly starts. If regular depreciation is claimed when the animal reaches maturity, the 20% allowance should be claimed in the first year that the regular depreciation is taken.

Useful Life

The Internal Revenue Service has established suggested guidelines for the useful life of livestock.

Animals	Useful Life
Cattle (breeding or dairy)	7
Horses (breeding or work)	10
Hogs (breeding)	3
Sheep and goats (breeding)	5

If shorter lives are used, the farmer must be able to prove why the animals are short-lived. Useful life, of course, relates to one's own practice, not to the intrinsic life; but if animals are usually disposed of before the end of their intrinsic lives, the estimated salvage value increases. If the taxpayer adopts ADR, he may use any life 20% longer or shorter than those set out without challenge.

No guidelines are offered for other animals such as race horses and fur-bearing animals. The useful life must be determined according to the particular facts and circumstances. The taxpayer may not use ADR unless there is a class life. Thus, if the taxpayer adopts ADR for these assets without a class life, he must apply to the Office of Industrial Economics in the Internal Revenue Service for assignment of a class life.

Salvage Value

The rule that salvage value may be reduced by 10% of the basis of the asset (see § 5.7) does not apply to livestock; the "true" salvage value must be used. Salvage value on livestock is difficult to determine.

Agents are less inclined to argue salvage value since recapture was extended to livestock. Similarly, farmers and ranchers may be less eager for low values, although they still have the use of tax money until it is paid back through recapture. Also if a taxpayer adopts the ADR system, his salvage value may not be adjusted unless it varies from the proper salvage value by more than 10% of the unadjusted basis of the property. See § 5.7.

Investment Credit

Livestock (other than horses) qualify for the investment credit. The credit does not apply to any livestock acquired if **substantially identical** livestock are disposed of within six months **before or after** the acquisition without a recapture of investment credit. See § 5.9.

Involuntary Conversions

An involuntary conversion occurs when an animal is converted into other property or money as a result of destruction in whole or in part, theft, seizure or requisition, or condemnation or sale because of disease or drought. See § 3.24.

Gains are not recognized if the proceeds are used to buy livestock similar or related in service or use. If the entire proceeds are not reinvested, gain is recognized to the extent that the amount realized from the conversion exceeds the cost of the animals bought as replacements. These rules apply to animals held for sale or for use in the business.

The postponement provisions do not apply to losses. Many losses will be fully deductible. Losses on animals used in the business resulting from involuntary conversions will offset gains from business assets if they are draft, breeding dairy, or sporting animals held twelve (or 24) months or more or other business animals held for more than six months in 1976, nine months in 1977 and one year in 1978. These losses offset gains which would have been treated as capital gains.

The special rule on casualties applies to livestock. Gains and losses from destruction or theft of assets used in the business are aggregated. A net loss is fully deductible and a net gain is placed in the hotchpot of gains and losses from property used in the business. See § 3.12.

The investment credit will be recaptured if the conversion is an "early disposition," but a credit may be taken on the new animals acquired as a replacement.

Diseased Livestock

Livestock destroyed by disease or on account of disease or livestock sold or exchanged because of disease are treated as if they were involuntarily converted. Insurance proceeds, sale proceeds or amounts received from the government on condemnation of diseased livestock may be reinvested in like animals, and the farmer may elect to have the recognition of the gain postponed.

An early disposition of livestock used in the business will result in a recapture of the investment credit, but replacement property may be subject to the credit. Acquisitions of used property may not qualify for the credit if they exceed the $100,000 ($50,000 after 1980) limit.

Livestock Sold Because of Drought

Similar treatment is given to a farmer who sells draft, breeding, or dairy livestock (regardless of how long they are held) because of drought. No gain is recognized if the proceeds are reinvested in similar animals before the end of the next two tax years. Special permission can be obtained to extend the replacement deadline. If the farmer does not replace the animals or if there is a loss, regular gain-and-loss rules apply.

Again the investment credit will be recaptured if the disposition is early, but acquisitions may be subject to the credit. Acquisitions of used property may not qualify for the credit if they exceed the $100,000 ($50,000 after 1980) limit.

Tax-Free Exchange

The only voluntary circumstance under which a farmer may avoid the recognition of gain or loss is on an exchange of business property or investment property for property of a like kind. In effect, the new property acquired on the exchange is treated as a continuation of the original property in substantially the same form. For example, the trade of a dairy cow for a dairy cow is a continuation of the original investment upon which neither gain nor loss would be recognized. The fact that the value of the animal received is more or less than the basis of the animal exchanged does not matter if the exchange qualifies as "tax-free." See § 3.27.

To qualify for a "tax-free" exchange, the animal must be held for productive use in the trade or business, for example, a breeding animal, or held for the production of income, for example, a race horse, and must be exchanged for property of a like kind which is held for the same purpose. Property is of a like kind if it is of the same nature or character although not necessarily of the same grade or quality.

It is clear that it is a "like kind" exchange if the trade is cows for cows, ewes for ewes, sows for sows, or bulls for bulls. Livestock of different sexes are not property of a "like kind."

Nearly the same result as that achieved by the tax-free exchange of heifers and bulls can be accomplished by a taxable trade of animals at a very young age. For example, an animal might be traded at age one month at a $25 value recognizing $25 of income. The $25 would then be the basis for the animal received on the trade. The ordinary income is minimal.

Again the investment credit will be recaptured if the disposition is early, but acquisitions may be subject to the credit. Acquisitions of used property may not qualify for the credit if they exceed the $100,000 ($50,000 after 1980) limit.

§ 2.4-2 Understanding the Futures Market

The speculator in futures has capital gains or losses. Hedging produces ordinary income or loss.

Commodity and livestock futures purchased for speculative purposes are capital assets. Ordinarily, speculating in grain, carcass or live cattle or hog futures results in a capital gain or loss.

However, a farmer engaging in a true hedging transaction as a protection against price fluctuations has ordinary income or loss from such transactions. This means that gains are fully taxable and losses fully deductible. See § 3.21.

If prices go up, the hedger may end up selling for substantially less than if he stayed out of the futures market. However, the purpose of hedging is not to get the best possible price, but to fix the price at an acceptable level in advance of sale. If today's futures quote looks good enough to allow reasonable profits and gambling on higher prices looks too risky, a hedge will assure that price.

Margin Calls

A price rise can be a real hazard if the speculator does not have the ready cash or an arrangement with a lender to meet "margin calls."

When first buying or selling futures, a "margin" or "good faith money" deposit is made with the broker. For example, to place a hedge on 100 head of cattle in his feedlot, the farmer would sell three live cattle futures contracts (40,000 pounds each) on the Chicago Mercantile Exchange. When the transaction is made, a deposit per contract must be deposited with the Exchange by the broker. The deposit may vary as the market fluctuates.

The broker, in turn, will ask for a deposit with him of enough margin to cover his deposit. He may ask for additional margin above that required by the Exchange, depending on the broker and the commodity being traded. Some brokers require a minimum account from a customer. Thus, it may not be practical to buy or sell one or two futures contracts.

Once deposited, the margin must be kept up to date if it is to continue to provide about the same degree of protection to the Exchange and the broker. This is one of the most misunderstood points of futures trading. Assume that the value of the October beef cattle future rises one cent per pound. As each futures contract for cattle is for 40,000 pounds, the value of the three contracts increases $1200.

If a farmer is holding such a contract, his margin would be reduced and his broker's protection would be reduced.

Brokers' practices vary, but in most cases, when a price change reduces the margin by 25% to 30% the broker will call for an additional deposit to make up the deficit. A farmer would be asked for an additional deposit. He must be ready to put a check in the mail immediately. If the broker does not receive the money within about two days, one or more of the contracts will be closed out. Should the price continue to increase, the holder of the contract will be called on regularly to bring his deposit back up to the required minimum.

Overselling

Suppose a farmer sold three hog futures contracts (30,000 pounds live weight each) which are equivalent of about 400 head of 225-pound hogs, at farrowing time, but he actually produced only 380 market hogs. He will have to buy back the equivalent of twenty more hogs than he has to sell on the cash market. If the market has gone down since he sold the contract, he will make a speculative profit on the twenty, but if it has gone up, he will lose money.

Unfavorable Basis Change

Basis (in the futures market as distinguished from tax basis) is the difference between the cash market price and the futures price. Suppose the farmer sold live cattle futures at $26 on the assumption that cattle at the local market would sell for $1 under the Chicago future. He figured on getting $25 for his cattle, less costs. It may turn out, however, that the basis is more than $1 when he sells; thus he gets less than the expected $25.

A futures contract is made for a certain grade and class of commodity. If his product does not match the description, the farmer takes another form of basis risk. He may, for example, hedge heifers against choice steer futures, figuring a $1 per cwt. spread between steers and heifers at delivery time. If the spread becomes greater, the hedge will be imperfect and he could lose money.

§ 2.5 TAX HANDLING OF MACHINERY

§ 2.5-1 Recovery of Investment

One of the major continuous activities in farm operations is the acquisition and disposition of equipment used in farming. The economics of cost, adjustment of equipment to particular needs, and the problems of the "over-machined" farmer are well-known aspects of farm management. But of equal importance to farm management and sound planning are the tax aspects of the handling of equipment.

The great tax benefit—the investment credit—was repealed for a period commencing April 18, 1969, but it was brought back for orders after March 31, 1971.

Depreciation deductions represent recovery of cost; recapture of the investment credit and depreciation means payment of a tax penalty. Sales of equipment can result in ordinary income or capital gains. Trading equipment may avoid gain completely. In many instances the dollar saved by the tax conscious management of equipment acquisitions and dispositions will produce a net return equal to many dollars of increased sales or decreased expenses.

The investment credit was repealed as of April 19, 1969. It was brought back for orders after March 31, 1971, and deliveries after August 15, 1971. There were a few transitional exceptions, notably property to be acquired under a contract binding on April 18, 1969; the others might apply to farmers, although it is doubtful. There are also a number of rules applicable to carry-overs and carry-backs of the credit. See § 5.9 for details.

Ordinary Depreciation

Depreciation is the planning tool which determines how the cost of a piece of equipment shall be recovered against the income tax system. Conservative management would try to allocate the cost of a machine in rough proportion to its annual use or decline in value. Some managers try to recover cost as fast as possible.

The tax law permits considerable management choice. While it has been common to spread the cost of a machine equally over the years of its useful life, there are faster methods of recovery of cost. An extra 20% first-year allowance and "fast" depreciation which recovers about two-thirds of the cost in the first half of the life may be claimed on many pieces of farm equipment. See §§ 5.2, 5.5.

Except when ADR is chosen, the Internal Revenue Service tries to make farmers maintain a reasonable relationship between tax write-off and actual replacement policies. Arguments arise over too short useful life and too low salvage value. Again ADR may be helpful on the salvage question.

In addition there is a very substantial penalty for too fast write-off if the asset is sold. The recapture provisions (see § 5.9) will treat as ordinary income, not capital gain, any part of the gain on the sale of equipment which represents depreciation previously taken.

Selecting Useful Life

Except where the farmer uses the ADR system, both useful life and salvage value are based on the individual farmer's normal period of use of a machine, not on its intrinsic life or the junk value at the end of that life. For example, if a farmer regularly uses a tractor for four years, the useful life to him is four years even though such a tractor has an inherent useful life of ten or twelve years.

Useful life is a factor in many determinations. Obviously, it sets the rate of depreciation. Fast depreciation is allowed only on property having a useful life of three years or more and the extra 20% first-year depreciation may be taken only on assets with at least a six-year life.

The Treasury has established a suggested ten-year useful life for all farm machinery. A useful life shorter than ten years may be used if it can be proved reasonable, but unless the actual life fits the Service policy, it will be difficult. See § 5.6.

If the farmer has elected to use ADR, however, he may use any period falling between eight and twelve years and know that he will not be questioned about useful life. Further, unless his estimate of salvage value is off more than 10% of the unadjusted basis, a Revenue Agent would not be authorized to change the salvage value.

Determining Salvage Value

Salvage value is the amount expected on the sale of a machine after it is no longer useful in the business. The aggregate of depreciation deductions plus the salvage value at the end of the useful life should equal the original cost or other basis of the property. A machine may not be depreciated below its reasonable salvage value, even under the declining-balance method in which salvage is not considered in setting rate or basis.

A minimum salvage value should be recognized on all assets, unless it is certain that a particular asset will be totally consumed. In estimating salvage value, there is some leeway on assets with a life of three years or more. The salvage value may be reduced by 10% of the basis of the asset. Thus, if a tractor cost $6000, had a useful life of eight years and a salvage value of $1000, salvage value may be reduced by $600 (10% of the basis of $6000) and $400 used as salvage value in computing depreciation. There is a further leeway of 10% if the ADR system is used by the taxpayer. See § 5.7.

Depreciation Techniques

The straight-line method of depreciation is the simple and common form of depreciation which will be used unless a farmer elects the more rapid write-off discussed in the text below. Under straight-line depreciation, the cost basis of a machine (less its estimated salvage value in excess of 10% of cost) is recovered in equal annual installments over its estimated useful life. If ADR is elected salvage value is ignored in setting depreciable rates. The property may not, however, be depreciated below salvage value even though it is ignored in determining annual depreciation allowances. See § 5.5.

For the farm manager who desires to recover his cost at a rate faster than the usual straight-line depreciation, there are two "fast write-off" techniques. The first is "bonus" depreciation. He may elect to claim extra depreciation in the year of purchase of a machine up to 20% of its cost. The remaining 80% of the cost is then used as the basis for regular depreciation using the straight-line method or one of the following faster methods.

The other technique is accelerated depreciation. A farmer may elect to depreciate a machine under depreciation methods which recover about two-thirds of the cost of the machine in the first half of its useful life. There are some details to these methods that warrant careful planning.

Extra First-Year Depreciation

As an additional fast write-off of part of the investment in depreciable equipment, up to 20% of cost is allowed in the year of purchase. The election applies only to tangible depreciable personal property purchased for use in the farm business. The property may be new or used but it must have a useful life of at least six years.

Equipment, machinery and movable structures that do not become part of the real estate qualify, but not buildings, structures or improvements to real property.

The equipment must be acquired by purchase or exchange. The extra depreciation allowance does not apply to property received by gift, bequest or inheritance. Property acquired from a related seller is not eligible. A machine bought from a member of the immediate family or from a controlled corporation or partnership will not qualify for the extra depreciation. Purchases from brothers or sisters are permissible.

The 20% depreciation applies to a machine received on a trade-in only to the extent of the cash paid on the trade-in. Sale of the old property and purchase of new rather than trading will be necessary if the full deduction is to be used. However, see the warning on phony sales and purchases at § 3.27.

The deduction of 20% of cost is to be taken in the first taxable year in which the property may be depreciated. If items bought late in the current year are held for the next year's depreciation schedule, the same pattern is followed in taking the 20% write-off.

Although the election is made each year for each machine acquired in that year, the deduction is limited to $10,000 worth of property per year on a separate return ($2000 deduction) and to $20,000 worth of property per year on a joint return ($4000 deduction). See § 5.2.

First Year Conventions

Under ADR a taxpayer may choose one of two conventions for the time property is placed in service. Under the "modified half-year convention," all property placed in service during the first half of the year is treated as placed in service on the first day of the year. Property bought in the last half of the year is not subject to depreciation until the first day of the next year. The other convention, "the half-year convention," treats all acquisitions as if made on the first day of the second half of the year. The convention used is applicable to all acquisitions of the year, but the taxpayer may make a new choice each year.

Fast Depreciation on New Property

The two common methods of fast depreciation are the double declining-balance method and the sum-of-digits method. These may be used only on new property with a useful life of three years or more. The useful life of the asset is related to the particular taxpayer rather than to the useful life of the asset as such. For example, if a farmer trades his pick-up truck every two years, the useful life is two years and the fast depreciation methods would not apply.

Either method of depreciation may be adopted for a piece of equipment the first time that it appears on the income tax return.

The machine must be new, that is, the original use must commence with the taxpayer. Fast methods may not be continued or started on any property in the hands of any person who acquired it from the original owner. For example, if an individual started the double declining-balance or sum-of-digits method on property and then contributed the property to a partnership, the partnership could not continue the fast method because it did not have the original use.

These fast depreciation techniques recover much more cost the first year and a much higher percentage of cost during the early years of the use of the property than the ordinary straight-line method.

The declining-balance method doubles the depreciation deduction in the first year. On a machine with a useful life of ten years, over 40% of the cost will be recovered in the first quarter of the life of the asset, and in the first half of the useful life about two-thirds of the cost.

The declining-balance method leaves an undepreciated residue at the end of the depreciation period. If the machine has not been depreciated to its salvage value, straight-line depreciation may then be used.

The sum-of-digits method is the other form of fast depreciation which gives higher deductions in the earlier years of useful life. The depreciation allowance is computed by applying a declining fraction to the cost of the property (reduced by salvage value if ADR is not elected).

The depreciation deduction under the sum-of-digits method will not be as great in the first year as under the double declining-balance method but the second-year deduction will be greater under the sum-of-digits method. Results at one-fourth and one-half the useful life of the asset will be substantially the same as the declining-balance method.

Fast Depreciation on Used Property

The very fast methods above apply only to new property but there is a more limited form of the declining-balance method which can be applied to used property. The rate is limited to 1 1/2 times (or 150% of) the straight-line rate as contrasted with twice (or 200% of) the straight-line rate for new property. The 1 1/2 declining balance may be elected for any piece of property for the first year in which the property is subject to a depreciation allowance.

Salvage value problems are postponed because any declining-balance method leaves an undepreciated residue and the question of whether this is proper salvage value cannot arise until the last year or years of life. Use of the ADR system results in a similar treatment for salvage value under all other methods of depreciation.

See §§ 5.3, 5.5 for explanation of depreciation methods and comparison of deductions allowed.

§ 2.5-2 Benefits and Traps on Dispositions

Sales and other dispositions of machinery and equipment may result in capital gains (only half taxed) or ordinary losses (fully deductible). This arises from the peculiarity of Section 1231 of the Internal Revenue Code which aggregates the gain or loss on a number of specialized transactions and treats a net gain as capital gain and a net loss as an ordinary loss. See § 3.12.

In addition to the gains and losses from dispositions of farm machinery and equipment, Section 1231 includes in its aggregate the gains and losses from sales, exchanges and involuntary conversions of cattle and horses owned for 24 months (other livestock owned for twelve months) and held for draft, breeding, dairy, or sporting purposes, farmland held more than six months (nine months in 1977; one year in 1978 and later years), timber sales under royalty contracts, the cutting of timber and unharvested crops sold with the land and the gains and losses from involuntary conversions of capital assets held more than six months (nine months in 1977; one year in 1978 and later years).

If the result of aggregating all Section 1231 transactions is a net gain, the gain is a capital gain. If the result of the aggregation of all of the Section 1231 transactions is a loss, then each asset is treated as producing ordinary gain or loss. As a practical matter, a net loss from the aggregation of Section 1231 transactions will be fully deductible.

Thus, it is an obvious advantage to try to realize Section 1231 gains and losses in different years so that gains will be capital gains and losses will be deductible in full.

A special rule, for casualty and theft losses, requires a preliminary netting process. Gains and losses from casualty and theft are first netted separately. If the net is a gain, the gains and losses fall under Section 1231. However, if the net is a loss, none of the casualty and theft gains and losses are included with Section 1231 transactions; in effect, there will be a fully deductible net loss from casualty and theft.

There are just two requirements to be met in qualifying machinery and equipment for Section 1231 treatment—(1) that it be used in the farm business, and (2) that it be owned more than six months in 1976, nine months in 1977 and one year thereafter. Thus, virtually all machinery and equipment is entitled to the special treatment.

Planning Dispositions of Property

In planning the disposition of a machine or a piece of equipment three tax traps must be carefully considered. First, if a machine which qualified for the investment credit is sold before the end of the useful life upon which the credit was computed, there will be an addition to tax in the year of sale.

Second, if a machine is sold at a gain, the part of the gain that represents a recovery of depreciation deductions will be taxed as ordinary income not capital gain.

Third, if the farmer has an "excess deductions account," the gains from machinery and equipment are subject to recapture as ordinary income to the extent of the balance in the account. Additions to the excess deductions account will not be made in years after 1975. However, an existing EDA must still be consumed by converting Section 1231 gain to ordinary income or through the earning of enough farm income to offset the balance in the EDA.

Investment Credit Recapture

If equipment for which a credit was allowable is disposed of before the end of its useful life, the tax for the year of disposition is increased by the amount of credit not fully "earned." This is the excess of the credit originally allowed over the amount of credit that would have been allowed if the estimated useful life had been the actual useful life, that is, the period from the date of placing the asset in service to the date of disposition. The same thing happens if, for example, the property is traded or exchanged (even in a "tax-free" trade) or is given away or converted to personal use or if a partner sells his interest in a partnership which holds such property.

There is no "disposition" of an asset, and no increase in tax is required, if the asset is transferred by death of the owner or of a joint tenant, or if there is a mere change in the form of doing business and the farmer retains a substantial interest in the business.

If an asset is destroyed by casualty, there is recapture. Replacement property may, however, qualify for the credit. See § 5.9.

Depreciation Recapture

Amounts which would otherwise be capital gains are converted into ordinary income on the sale of depreciable property if the gain represents the recovery of depreciation. Any gain in excess of the amount attributable to a recovery of depreciation is treated as capital gain or aggregated with Section 1231 items. See § 5.10.

This rule covers all depreciable machinery and equipment which is or has been subject to depreciation. There is no useful-life test.

In addition to machinery and equipment, recapture takes place on other tangible property (not including a building or its structural components) which was depreciable while used in the farming business or as a research or storage facility in connection with the farm business. These are primarily land improvements and are discussed in § 5.10.

If the machine is sold on the installment method the gain in each installment is treated as ordinary income until all of the depreciation is recaptured; thereafter, any capital or Section 1231 gain is reported.

There is no recapture by the donor on a lifetime gift. But the donee will report all of the ordinary income recapture (including depreciation taken by the donor) when he disposes of the property. Nor is there any recapture by a decedent on a transfer at death. However, the estate, heir, legatee or other recipient receives a carry-over basis on death and there will be carry-over of the recapture that would have been reported by the decedent. See § 5.10.

If a machine is disposed of in a tax-free exchange or by involuntary conversion, the amount of ordinary income to be reported generally cannot exceed the gain actually recognized on the transaction. Any unrecaptured depreciation is carried over into the replacement machine.

If depreciable property is donated to a charitable organization, the amount of the deduction is reduced by the amount which would have been treated as ordinary income if the property had been sold for its fair market value. See § 5.10.

Trades and Reinvestments

A farmer does not report gain or loss on a trade of farm equipment for property of a like kind. The equipment acquired on the trade is treated as a continuation of the original equipment. For example, the trade of a tractor for a tractor is a continuation of the original investment upon which neither gain nor loss would be recognized. The fact that the value of the tractor received is more or less than the basis of the tractor exchanged does not matter.

Generally recapture of depreciation is also postponed until the acquired asset is disposed of. However, any unearned investment credit on the property traded is recaptured even on a "tax-free" exchange.

Property must be held for productive use in the farm business and must be exchanged for like property which is held for the same purpose. Property of a like kind means of the same nature or character although not necessarily of the same grade or quality. For example, a tractor may be traded for a tractor or a truck, but not a corn picker for a boat, or a tractor for ten acres of land.

If the trade is solely one piece of equipment for another of a like kind, no gain or loss is recognized and the new property takes over the basis of the old. If cash boot is paid on the trade there is no gain or loss recognized and the basis of the acquired equipment is the basis of the old plus the cash paid.

If, on an exchange of like kind property, the taxpayer also receives cash or other additional property, he must recognize the gain on the transaction to the extent of the cash and the fair market value of the additional property.

Upon disposition of the property acquired on an exchange, the postponed gain or loss will usually be reported. If property passes to a new owner at death, it takes a carry-over basis and some or all of the postponed gain or loss will be recognized.

Often, farm managers want to report gains and losses on substitutions of equipment. Because most farm equipment falls under Section 1231, gains can be capital gains and losses can be ordinary losses. If the full gain or loss is recognized on the disposition of the old machine, the basis of the new machine is its cost. There is an obvious tax advantage if the gain recognized on the old machine can be a capital gain and there is a new higher basis for depreciation on the new machine which offsets ordinary income. Similarly, if a loss can be taken against other income, there is a greater present tax benefit than burying the loss in the basis of the new property.

Two factors conspire against the recognition of capital gain or ordinary loss. First, the gain may be ordinary income because it represents recapture of depreciation. Second, the Internal Revenue Service is suspicious of the separation of the sale of old equipment from the purchase of new to avoid an "exchange" of the old equipment for the new and to obtain capital gains or a loss deduction.

The Service says that in the usual case of the separate sale to and purchase from the local dealer, the transactions are reciprocal and mutually dependent. If the acquisition of the new equipment from the dealer is contingent on the dealer's accepting the used equipment and granting an allowance equal to or in excess of its fair market value, the deal is really a trade-in.

See § 3.27.

Reinvestment After Involuntary Conversion

There is an involuntary conversion when money is received for equipment as a result of its destruction, in whole or in part, theft, seizure or requisition or condemnation. Insurance proceeds or other compensation in excess of the adjusted basis of the property destroyed, stolen, or condemned will produce a tax gain. If no compensation is received or if the compensation is less than the basis of the property, there will be a loss.

If money is received as a result of an involuntary conversion, gain will be recognized unless a special election is made. A farmer may elect not to recognize the gain if he purchases, within the replacement period, similar property the cost of which equals or exceeds the proceeds from the old property. If the cost of the replacement property is less than the proceeds received, gain will be recognized to the extent of the unexpended proceeds.

Losses on involuntary conversions are recognized. The postponement elections do not apply to losses.

The replacement must be made within two years after the close of the first taxable year in which any part of the gain upon the conversion is realized except for business or investment real estate subject to condemnation. The taxpayer has three years to replace it. Upon application to the District Director, an extension of either of these periods may be granted.

The replacement property must be similar to the property lost or destroyed. The physical characteristics and end uses of the converted and replacement properties must be substantially the same.

Under ordinary circumstances, a farmer who has lost a tractor will replace it with a tractor. The cautious rule is to replace with a machine that performs the same function. It may be bigger or smaller or of a different brand. But one would not consider a corn picker similar to a tractor, or a field wagon similar to a motor truck. See § 3.24.

Losses on Machinery

All losses on farm machinery and equipment will be deducted but the loss may be deductible in full or it may merely offset gains that would otherwise be capital gains. In addition to losses which might be incurred on sale or other dispositions, losses on farm machinery and equipment may result from casualty, theft, abandonment or retirement.

A casualty loss is the complete or partial destruction of property resulting from a sudden, unexpected or unusual force. Theft is the criminal appropriation of property.

There is no current loss deduction because of deterioration or non-physical influence. Future losses or depreciation of property are not currently deductible. The damage must be of a physical nature resulting from accident, fire, storm, flood or other sudden destructive force. See § 4.17.

A casualty or theft loss of property held for business (or even for personal) purposes becomes a part of the net operating loss which may be carried back for three years and forward for seven years. See § 4.17.

If the usefulness of an asset used in the business suddenly terminates because of changed conditions and the asset is discarded permanently, a deductible loss results.

Such a loss requires proof of the unforeseen circumstance causing the property to be prematurely discarded. Since it is usually difficult to prove, a careful farmer will ordinarily sell machinery or equipment as junk so that the loss can be accurately determined. See § 4.19.

§ 2.6 TAX PLANNING FOR SELLER AND BUYER OF A FARM

§ 2.6-1 Tax Conflicts of Seller and Buyer

This and the following sections consider the tax aspects of the sale or exchange of a farm, the immediate problems of the new

owner of the farm and some of the continuing tax considerations of the new owner as he plans the future of his farm operation.

For tax purposes, the sale of a farm should be thought of as the sale of separate assets, namely (1) land; (2) depreciable real estate, such as buildings, fences and tiling; and (3) house and land site. Even though a farm is sold as a package for a lump sum, the sale must be broken down. Sales price and basis must be determined for each item.

The land is nondepreciable so the tax basis is the original cost if purchased. If the land was inherited, the tax basis is the fair market value at the time of the death of a decedent who died before 1977. If he dies thereafter, his own basis carries over to the estate and may be adjusted for appreciation before 1977 and for state and federal death taxes on increases in value after January 1, 1977. If acquired by gift, the basis is generally the same as it was in the hands of the donor. See § 5.8.

There might be some capitalized items such as earthen dams and terraces which increased the basis if they were not written off as soil and water conservation expenses. See § 5.8. Land-clearing expenses which were not deducted increase the basis. See § 4.14.

The tax basis of the buildings and other depreciable items is usually the cost or other basis when acquired, plus the improvements less depreciation. The house is a capital asset which would have an original basis plus improvements. The house is not depreciable except for that portion used in the farm business. If the seller has a tenant farmer occupying the house, it is a depreciable farm building.

The figure obtained from all of the additions and deductions to the basis from the time the property was originally acquired gives the tax basis on disposal. The difference between the tax basis and the selling or exchange price is a gain or a loss.

If the farm was owned for over six months in 1976, nine months in 1977, and one year thereafter, the gain will be a capital gain, except for the portion of the gain that represents a recapture of depreciation, land deductions or "excess deductions account." There could also be a recapture of investment credit, for example, on fences, certain storage facilities, tiling, and silos. See § 5.9. If a growing crop is sold with the land, the crops usually produce capital gains. See § 3.16.

The seller may want to minimize tax by spreading the gain through an installment sale. See § 3.26. Or he may be forced to sell because of condemnation and he wants to acquire another farm without paying tax on the gain on the condemned property. See § 3.25. Perhaps he plans to acquire another farm and wants to avoid any gain on the mere change of investment. See § 3.27.

On the sale of the residence portion of the farm, a farmer 65 or over can avoid gain on the first $35,000 ($20,000 in years before 1977) of sales price and any farmer can reinvest the sales price of the farm residence in a new home and avoid tax on the gain. See § 3.28.

Buyer's Standpoint

The buyer who is considering his future farming career does not buy a "farm" as an entity but (1) land; (2) land improvements; (3) buildings; and (4) house and land site. He must allocate his purchase price to the various assets so that he has a basis for future depreciation and sale. See Chapter 5.

As a new owner, he will be concerned about the deductibility of immediate expenses in the development and improvement of the farm. See §§ 4.12 to 4.14.

§ 2.6-2 Planning the Sale of a Farm

When planning to sell a farm, the transaction must be examined not as the sale of an integrated unit, but as the sale of a bundle of assets, some of which will have to be separated for special tax treatment. The seller may seek to avoid gain by trading the farm for other real estate. See § 3.27.

The gain or loss on a sale will be reported in the year of sale unless the seller spreads the gain over a period of years by electing to use the installment sale method. See § 3.26.

A loss on the sale of a farm is treated as a farm operating loss. Any losses on the sale of business assets (depreciable property and real estate) are business losses and become part of a net operating loss. See § 4.18.

A seller does not sell a farm as a unit; he sells land, buildings, residence, land improvements, fences and other fixtures attached to the land and growing crops. There are particular tax aspects to each of these.

The sale of a farm results in a taxable gain or loss even though the proceeds of the sale are immediately invested in another farm. There are only two circumstances in which a farmer may shift his investment from one farm to another and not recognize the gain on the old farm: 1) If the farm is condemned, he may postpone the taxable gain if he reinvests the condemnation proceeds—see § 3.25; 2) If the farm is exchanged for another farm (not sold for money), the gain or loss on the old farm is not currently recognized—see § 3.27.

One other provision postpones the gain on the farm residence (house and land site) if the farmer buys another residence—see § 3.28.

Land and Buildings

Land sales fall under Section 1231 of the Internal Revenue Code which aggregates gains and losses from dispositions of business assets and certain other items. The basis of the land is the portion of the original cost of the farm allocated to the land plus the cost of land improvements which were not deductible or which are not separated for special tax handling. See § 4.12 on developing the farm for illustrations of such capitalized costs.

Buildings were separately treated on the purchase of the farm because they are depreciable property and a part of the original cost was allocated to each of them. Subsequent structures have a separate cost basis. The basis for each structure is reduced by depreciation that was claimed or should have been claimed. The gain or loss on the buildings will also fall under Section 1231. If fast depreciation was claimed, and the building is sold at a gain, some of the depreciation on buildings may be recovered as ordinary income. See § 5.10.

The sale of land at a gain may also require the recapture as ordinary income of some or all of the deductions previously taken for soil and water conservation expenses and land-clearing expenses. See §§ 5.12, 5.13.

The residence must be carved out of the sale for two reasons. First, if there is a loss on the residence it may not be deducted or set off against the gain on other portions of the farm. Second, if the seller plans to replace the residence without recognition of gain or claim the exclusion from gain allowed to persons 65 or over, he must have a sales price for the residence. See § 3.28. On the original purchase of the farm, some part of the original cost would have been allocated to the residence. This plus the cost of improvements will be the basis for gain or loss.

Other Improvements

Land improvements and fixtures raise special problems. If the land improvements involved no tangible depreciable property, they can be added to the basis of the land as suggested above. However, tangible depreciable property, even though affixed to the land, is eligible for the investment credit, and if the three-to seven-year period of the credit has not expired, there will be a recapture of the credit as additional tax. See § 5.9. Similarly, such property is subject to recapture of depreciation if the item is sold at a gain. Tiling, silos, storage facilities, feeding floors, water systems, well casings, etc., will require separate determination of gain or loss. See § 5.10.

As a practical matter, growing crops sold with the land do not cause difficult tax problems for the seller. He cannot deduct the cost of raising the crops. The costs are added to the basis of the land, and the price of the crops is included in the price of the land. But the buyer needs a separate allocation of the purchase price of the crops. See § 3.16.

A special problem arises for a farmer who has had $50,000 of nonfarm income in prior years and had to maintain an "excess deductions account." If there is a balance in the account, the gain on the sale of farm assets which would ordinarily be capital gain will be recaptured as ordinary income to the extent of the EDA balance. See § 5.13.

It is clear that a lump-sum sales price or a price in dollars per acre is not realistic for tax purposes. The price must be allocated to the assets sold to the extent that these require separate handling, so that gain or loss can be determined on each item.

Taxes and Interest

Real estate taxes for the year of sale are split between buyer and seller in proportion to the number of days in the real property tax year that each owned the property. If the buyer pays all of the taxes, the seller's prorata share of the taxes is additional purchase price received. If the buyer's portion is paid by the seller, the purchase price is reduced by that amount. Each deducts his own share of the taxes for the year of sale. See § 4.9.

Deferred payments of the sale price must carry at least 6% simple interest or else an interest element will be carved out of the sales price at a 7% rate. This, of course, means that the seller will report ordinary interest income whether he provides for it or not. See § 3.26. The technique of high purchase price and low or no interest which was not uncommon in prior years is no longer good tax planning.

Tax Savings

Farmers are often puzzled about the tax-saving methods available when disposing of farm property. This may stem from confusing the different rules that apply to an exchange, a sale of a residence, and a replacement after condemnation.

Some farmers think that they can sell a farm and reinvest in another farm within a specified period of time and thus avoid any tax consequences. This is not true unless the farm was condemned. See § 3.25. If a farmer sells his residence and buys a new one, he may avoid the gain on the sale of the old residence. See § 3.28. The only way to avoid gain on the substitution of one farm for another is to trade farms. See § 3.27. Gain can be spread over a period of years by use of an installment sale. See § 3.26.

Method of Payment

Farms are often sold under arrangements by which the payments of the purchase price are made over a period of years. Under the ordinary rule, the transaction is treated as complete in the year of sale and the gain or loss is recognized and reported in the year of sale. However a farm seller may elect to use the "installment method" and report as gain a part of each installment as the payments are received.

The amount realized on the sale of a farm is the cash plus the fair market value of the obligations of the purchaser. The sale is complete and the income is realized when the title or the burdens and benefits of ownership pass to the buyer. If the obligations of the buyer are worth their face value, the total gain is recognized in the year of sale even though all of the payments have not been received.

If the value of the obligations of the purchaser is less than face value, the gain in the year of sale is based upon the cash plus the value of the purchaser's obligations. When the obligations are ultimately paid off at the face amount, there will be additional income to report which will be ordinary income even though the gain on the original sale was capital gain. See § 3.26.

Installment Sale Election

To alleviate the hardship of paying tax on money not yet received, the seller may elect to report a sale on the installment method which defers the tax on the gain until the payments are received. A prorata part of each payment is treated as gain and the tax on that gain is reported in the year that the installment is received. Usually, the installment sale results in a lower over-all tax because of the spread of the gain over a period of years.

No payment is necessary in the year of the sale, but payments in the year of the sale may not exceed 30% of the selling price. An installment-sale contract may provide for nothing in the year of the sale and the balance in at least two payments in future years.

The amount returned as income from an installment sale is that proportion of the payments actually received in the year which the gross profit on the entire transaction bears to the total contract price.

The installment-sale election applies only to sales that produce an income tax gain. A loss on the sale of a farm may not be spread over the payment period. It is treated as a net operating loss which can be carried back three years and forward seven years.

Disposition of the installment obligations results in immediate recognition of all of the postponed gain. A sale, gift or other disposition or the satisfaction of an installment obligation causes all of the postponed gain to be immediately recognized.

The postponed gain is not recognized on a transfer by death of the rights to payments under the installment obligations. If a farmer dies owning installment obligations, the executor, administrator or legatee reports the income in the same manner as the decedent would have done if he had lived. The successor is entitled to an income tax deduction for death taxes paid on the installment obligations. See § 3.26.

Sales Must Bear Interest

A farmer cannot increase the purchase price of his farm and charge no interest or a very low interest to convert ordinary interest income to capital gain sales price. In such case, the interest element will be carved out of the sale price and treated as interest. Unstated interest even arises on a sale at a loss; the loss is increased by excluding the unstated interest from the proceeds.

If the contract is in excess of $3,000, and one or more payments are due more than a year after the date of sale, and the interest rate is below 6%, a portion of each deferred payment discounted at 7% is treated as interest. See § 3.26.

Unharvested Crops

Unharvested crops sold with the land are treated as part of the sale of the land and are entitled to capital gain or Section 1231 treatment. The land must have been held for more than six months in 1976, nine months in 1977 and one year thereafter. There is no rule on the age of the crop. The seller must have no right or option to reacquire the land directly or indirectly (except through a mortgage or other security contract).

Condemnation or seizure or sale under threat of condemnation is treated the same as a sale. Sale by a tenant of his leasehold with growing crops does not qualify for this capital gain treatment.

The seller must treat the costs of producing the crop as a capital investment rather than as an expense. In the year of sale of unharvested crops with the land, the farmer may not deduct the costs of raising the crop. These include all items attributable to the crop, including depreciation. For example, depreciation on machinery used in cultivating the crop must be capitalized if the entire farm and the growing crops are sold. These expenses are added to the basis of the land and reduce the gain or increase the loss. See § 3.16.

Personal Residence

A residence is a capital asset of the taxpayer which will ordinarily produce gain or loss when sold. Gain is capital gain and recognized unless the taxpayer reinvests the proceeds in another residence, or if he is 65 or over and elects to exclude the gain on the first $35,000 ($20,000 before 1977) of the sale price. A loss on the sale of a residence is a personal nondeductible loss. Only a casualty loss on a residence is deductible.

Replacement of Residence

One who sells or exchanges a residence at a gain is not taxed if he reinvests the proceeds in another residence within eighteen months or if he bought a new residence within eighteen months prior to the sale. Thus, if the new residence costs as much or more than the price received for the old, none of the gain is recognized for income tax purposes. If the new residence costs less, the gain is recognized to the extent of the difference between the sale price of the old and the cost of the new.

If the taxpayer builds a new home, he must complete it within eighteen months before, or two years after, the sale of his old residence, and construction must have begun before the expiration of eighteen months after the date of the sale of the old residence.

A farmer who sells his farm with the residence must allocate a portion of the purchase price and the original basis to the residence. If the old residence or the entire farm was sold on an installment sale, the recognized gain on the residence would be spread over the period of installment payments. See § 3.26.

Exclusion of Gain by Farmer 65 or Over

An individual aged 65 or over may exclude from income the capital gain attributable to the first $35,000 of the sale price of his personal residence. The property must have been owned and used by the taxpayer as his principal residence for at least five years out of the eight-year period before the sale.

If the sales price is $35,000 or less, all of the gain is excluded. If the sales price is over $35,000, the exclusion is limited to the part of the gain that is in the same ratio to the total gain as $35,000 is to the sales price. Before 1977, this provision applied to only the first $20,000 of sales price rather than the first $35,000.

On the sale of a farm, the residential portion is carved out. A residence condemned, destroyed or otherwise involuntarily converted is treated as sold, and the gain may be excluded.

If a taxpayer excludes part of the gain because he is 65 or over, the remainder of the gain will be postponed to the extent that he reinvests the sale proceeds in another residence. See § 3.28.

Seller's Tax Recapture

The seller of a farm faces several forms of tax recapture. The simplest one is the recapture of depreciation claimed on buildings in excess of straight-line depreciation. See § 5.11. More troublesome will be the possibility of recapture of the investment credit and recapture of depreciation on tangible depreciable property. See §§ 5.9, 5.10.

Neither the investment credit nor the recapture of ordinary depreciation applies to land or buildings but the definitions under both of those provisions include tangible property attached to the land.

In addition, some or all of the deductions for land-clearing and soil and water conservation expenses may be recaptured as ordinary income. In addition, any amount in the "excess deductions account" may be recaptured as ordinary income. See §§ 5.12, 5.13.

Nonbuilding Recapture

Both the investment credit recapture and depreciation apply to depreciable tangible property (other than a building or its structural components) if the property was used as an integral part of farm production or if it was a storage or research facility used in connection with farm production.

Any depreciable item used in the farm business which is not properly a building fits the definition of property subject to investment credit and depreciation recapture on sale. Fences, water installations, outside light and power systems, depreciable parts of dams and ponds, tiling, pavements, silos, grain bins, etc., are tangible and depreciable but are not buildings or components. Items of equipment that are attached to buildings but which are not part of the basic structure and can be removed without substantial damage, for example, fans, motors, ventilators and equipment used for environmental control, are in the same category.

Thus, on the sale of a farm, all of the items which were subject to the investment credit must be examined separately. If the item was not held for the life that was used in computing the investment credit, the "unearned" portion of the credit must be added to the tax in the year of sale.

Similarly, each piece of property attached to the real estate which is subject to depreciation recapture must be handled separately. Gain must be computed on each item to determine whether any depreciation is to be recaptured as ordinary income.

See §§ 5.9, 5.10 for the general background of the investment credit and Section 1245 depreciation recapture provisions.

Recapture on Buildings

The rules on recapture of depreciation on buildings are quite different from the rules on recapture of depreciation of personal property. Section 1250 of the Internal Revenue Code applies a special recapture theory to the buildings and structural components which were excluded from the nonbuilding recapture provisions. There are some special complications because there are different rules for post-1969 and pre-1970 depreciation.

With respect to pre-1970 depreciation, there are three important limitations on recapture. First, the depreciation actually recaptured is only the "additional depreciation" (the excess of fast depreciation over straight-line depreciation) claimed from 1964 through 1969. Second, if the property was held over twenty months, only a portion of the additional depreciation is recaptured. Third, if the property was held for ten years or more, there is no recapture at all. See § 5.11.

Under the law which applies to depreciation claimed in 1970 and thereafter, all of the depreciation in excess of straight line is recaptured. There will be two computations, one for the period before 1970 and one for the period after 1969.

Depreciation on real estate acquired after July 24, 1969 is limited to 150% declining balance on new buildings and to straight line on used buildings.

A farmer who always used straight-line depreciation on his buildings will not fall under this section at all, with one exception—if the property was held less than a year and there is a gain on sale, any depreciation claimed is converted into ordinary income to the extent of the gain.

Condemnation

If a farm is condemned or sold under threat of condemnation, the transaction is treated as an involuntary conversion and the gain or loss falls under Section 1231. Recognition of gain (but not loss) may be avoided to the extent that the proceeds are invested in other real estate.

Depreciation is recaptured only to the extent of the gain recognized. Any unrecaptured depreciation is assigned to the new property. Any unearned investment credit on tangible property attached to the real estate will be recaptured. See § 5.9.

Loss sustained from the condemnation of the residence portion is not deductible, but a gain will be taxable unless a new residence is acquired. The taxation of any gain on the residence may be postponed either as an involuntary conversion or under the sale and replacement of residence provisions. See §§ 3.25, 3.28.

Exchange of Farms

The ordinary rule is that gains and losses are recognized if property is exchanged for other property. The gain or loss is the difference between the value of the property received and the adjusted basis of the property given up. The gain or loss will be ordinary or capital depending upon whether the asset transferred was an ordinary asset or a capital asset or falls under Section 1231. There is one exception to this general rule.

A farmer may avoid the recognition of gain or loss on an exchange of farm property for property of a like kind. The property acquired on the exchange is treated as a continuation of the original investment into a new form. For example, the trade of a farm for a farm or other real estate is a continuation of the original investment upon which neither gain nor loss is recognized.

The property transferred must have been held for productive use in the farm business or for investment and must be exchanged for property of a like kind which is held for the same purpose. Property is of a like kind if it is of the same nature or character although not necessarily of the same grade or quality. For example, an improved farm may be traded for an unimproved farm or a farm may be traded for city real estate, but not ten acres of land for a truck.

Upon disposition of the property acquired on an exchange, the postponed gain or loss will usually be reported. If property passes to a new owner at death, it takes a basis equal to the old basis except that if gain is recognized the basis may be adjusted by increases in value before 1977 plus state and federal death taxes on increases in value after 1976. It is not uncommon for a farmer to trade his farm for an apartment building or other real estate to avoid the gain during his life, and then leave the property by will to a beneficiary who takes a basis under these rules which permits all the pre-1977 appreciation to go untaxed.

"Tax free" trades may be arranged among any number of parties. It is perfectly feasible for A, who owns a farm, B, who owns an office building and C, who owns an apartment, to arrange a three-way exchange so that each acquires the property he desires. Differences in value can be adjusted by cash payments. Gain will be recognized only by the person who receives cash and then only to the extent of the cash. If the properties are mortgaged, it is possible for each to assume the mortgage on the property he receives on the exchange. To the extent that one party is relieved of a greater liability than he assumes, the excess will be treated as cash received and he will recognize any gain up to the amount of the excess. See § 3.27.

§ 2.6-3 Planning the Purchase of a Farm

The preliminary planning for the buyer of a farm is somewhat similar to that of the seller, but it is more complicated and it has long-range effect. Generally, the buyer must allocate the purchase price to individual assets with much greater precision than the seller. He must assign a basis to each of the assets he buys. The allocation should follow respective market values and the allocations should be stated in the contract.

The buyer must allocate part of the purchase price to the land and he would be expected to be conservative because he will not recover this cost until he sells the land. If there is timber, or gravel, or mineral deposits, part of the land cost should be allocated to these. If the purchase included growing crops, allocation of part of the price to the crops is vital because the buyer will deduct this cost when he harvests and sells or feeds the crops.

There must be an allocation to each building and structure to provide a basis for depreciation and to each asset which qualifies for the investment credit.

The residence must be carved out so that it has a separate cost. It may be a replacement for an old residence sold at a gain or the farmer will be thinking ahead to possible replacement of this residence without gain or the exclusion of gain on sale of the residence when he is 65 or over. In allocating the purchase price, he faces the dilemma of attributing value to the residence to avoid future gain or to the depreciable buildings to provide a higher basis for depreciation.

Depreciation

The allocations of the total purchase price will establish depreciation bases for the individual assets. Then the rate of depreciation must be determined. A farmer cannot depreciate fast on the theory that capital gains lie ahead when he sells the property. On any tangible property, including a storage or research facility (not a building), depreciation recovered on sale will result in ordinary income. Thus, it is well to be cautious in setting depreciation rates on tiling, roadways, fences, feeding floors, silos, cribs, etc. Similarly, the excess depreciation on buildings will be recaptured.

See §§ 5.10, 5.11.

Investment Credit

The buyer needs an allocation of the purchase price to those items which qualify for the investment credit—for example, fences, tile drains, feeding floors, grain storage structures, etc. These must have an ascertainable "qualified investment," i.e., a cost. See § 5.11.

Development of New Farm

In the early stages of establishment of a new farm, some of the expenditures must be capitalized, and some may be capitalized or deducted at the new owner's option. Generally, if expenditures are capitalized, they will not be recovered until the farm is sold because they are added to the basis of the land which is not depreciable. Expenditures for timber, orchards, and groves are recovered through depletion on timber and depreciation on orchards and groves.

There are two definite periods in the life of a farm, orchard or ranch. These are the pre-productive period and the productive period.

During the pre-productive period, the farmer has the option to deduct or capitalize current ordinary and necessary business expenses. A farmer must care for his orchard, which does not yet produce income. He must cultivate and spray the trees, pay for the water he uses for irrigation, and buy fertilizer which has to be replaced each year. He has the option to deduct these ordinary and necessary business expenses or to capitalize them.

No deduction is allowed for costs of clearing brush, conditioning land, tree planting, drilling wells, building irrigation ditches, laying pipes, installing drain tile, constructing reservoirs or dams, etc. These are capital expenditures.

The special elections to deduct soil and water conservation expenses, and the cost of fertilizer and lime do not apply during the pre-productive period. However, expenses for clearing land to make it suitable for farming may be deducted but the limitation of 25% of taxable income from farming may make the election worthless. See §§ 4.13, 4.14.

When the farm, orchard or ranch reaches its productive period, the farmer loses any option he had during the pre-productive period. The farm is now a full-fledged operating business. Ordinary and necessary expenses cannot be capitalized. See § 4.12. A farm syndicate and some farm corporations may not be entitled to deduct these costs.

Citrus and Almond Groves

There is an important exception to these rules. Any amount attributable to the planting, cultivation, maintenance or development of a citrus or almond grove which is incurred before the end of the fourth taxable year beginning with the year in which the trees were planted must be capitalized. There is no election to deduct expenses during this period. This rule does not apply to replacements resulting from freeze, disease, drought, pests, or casualties. After 1975, this exception applies to all vineyards and orchards developed by "farm syndicates."

Conservation, Land-clearing, Fertilizer and Lime

Expenditures for soil and water conservation in respect of land used in farming or for the prevention of erosion of land used in farming are usually capital expenditures but a farmer may elect to deduct them. The election is made in the first year in which the farmer has any soil and water conservation expenditures. The deduction in any one year may not exceed 25% of the gross income derived from farming. Any amount in excess of the allowable deduction may be carried over and deducted in future years subject to the percentage limitation. See § 4.13.

Land-clearing expenditures are ordinarily capital costs of making the land suitable for use in farming. However, some or all of these may be deducted if the farmer so elects. The deduction is limited in any taxable year to the lesser of $5000 or 25% of the taxable income derived from farming. See § 4.14.

Another factor in the planning mix is the possibility of recapture of these items. Some or all of the deductions for soil and water conservation expenditures and land-clearing expenses will be recaptured as ordinary income against the gain from the sale of the land if the land is held for less than ten years. See §§ 5.12, 5.13.

Even though there is a long-term benefit, expenditures for fertilizer, lime and ground limestone, marl or other materials to enrich, neutralize or condition land used in farming and the cost of application may be deducted. This is an election to be made each year. See § 4.7. There is no recapture of this deduction.

Drilling Water Wells

Costs of drilling an agricultural well must be capitalized. Well equipment, pumps, pipes, etc., are subject to depreciation. Other drilling costs may be depreciated if the well has an ascertainable useful life—which is rare.

Amounts expended in the drilling of test holes are capital expenditures and are not deductible. If a sufficient supply of water is found and a water well drilled, abandonment of the test holes does not give rise to a deduction. The cost of the test holes is added to the cost of the producing well.

Demolition of Buildings

Demolition of buildings to make way for new structures may or may not result in a deductible loss. Sometimes a farmer acquires land with buildings on it and then demolishes the buildings. If he has no allowable loss, the adjusted basis of the buildings demolished and the costs of demolition are added to the basis of the new structure or the land.

If the land is acquired with intent to demolish the buildings, no deduction is allowed for loss on the buildings, and the cost of the buildings and the costs of demolition are added to the basis of the land. If the buildings are in fact used for a time, a portion of the basis of the property is allocated to the buildings for depreciation. The unrecovered portion of the basis so allocated is a loss on demolition or is added to the cost of the land. See § 4.19.

Other Capital Expenditures

Except for soil and water conservation expenses, land-clearing expenses and costs of fertilizer and lime, capital expenditures on farm land where depreciable materials are not used do not result in an offset against gross income until the farm is sold. These expenditures can be used by the taxpayer only to increase the basis of the farm for the purpose of computing gain or loss on subsequent sale or exchange. For exchange, Forest Service grazing permits and state grazing leases, costs of a survey to establish boundaries or ascertain topography, and legal fees to acquire title are nondeductible capital expenditures. See § 4.21.

§ 2.7 WISE TIMING OF TAX CHOICES

§ 2.7-1 Elections for the New Farmer

Several elections must be made by any taxpayer reporting income for the first time. Some elections may be made by existing taxpayers who start farming or who change their form of business organization, for example, to a partnership or a corporation. Generally, once an election is made, it may not be changed without the permission of the Commissioner of Internal Revenue.

Even though the accrual method is required for other taxpayers engaged in production and sale of goods, and even though the accrual method would more clearly reflect a farmer's income, individual farmers may use the cash basis. Most corporations and partnerships may also use the cash basis. However, corporations which are not 50% or more owned by one family, or which have not elected to have their income taxed to their shareholders, or which have had more than $1,000,000 in receipts in any year after 1975 may not use the cash method. If a partnership has a corporate partner which may not use the cash method, then the partnership is similarly prohibited. Such corporations and partnerships must also capitalize pre-productive expenses of most crops.

Most farmers selected either the cash or accrual method on the first farm return filed. However, a new election is made by a newly organized farm partnership or corporation unless it is required to use accrual accounting under the 1976 rules noted above. An election may be made for a newly purchased farm if it is a separate business.

A farmer may report his farm business income on the cash basis and other business income on the accrual basis, or his farm business on the accrual basis and his personal affairs on the cash basis.

Cash or Accrual Basis

Most farmers use the cash-receipts-and-disbursements basis. This means they can disregard inventories as a factor in computing income, account for income items when received, and claim deductions for expenses when the money is paid out.

Cash-basis accounting for the farmer is simpler, it is closer to his actual cash position, and produces more capital gains on sale of draft, dairy, breeding, and sporting animals raised.

Before 1977, neither the decedent nor his beneficiaries after his death paid any income tax on the crops and livestock on hand at death. Now, however, the decedent's basis will carry over to the estate and other successors, so this advantage of the cash method is pretty well gone.

The accrual method of accounting more clearly reflects the income from ordinary farming operations than the cash method because the difference between the beginning and closing inventories of the farmer affects the reportable income or loss for the year. Costs which have gone into products not yet sold are, in effect, carried over to be offset when the sale is made. Expenses for which the farmer is obligated, whether paid or not, business losses and depreciation are deducted from gross income.

A hybrid accounting system of any combination of methods is permitted if it clearly reflects income, for example, accrual of sales and purchases and cash accounting for other items. See § 6.1.

The accrual or hybrid basis farmer was not required to keep an "excess deductions account" to recover farm losses as ordinary income. He had to use inventories and capitalize all expenditures properly chargeable to capital account including those which he had an election to deduct. See § 5.13. But after 1975, even cash-basis taxpayers will not add any losses to their EDA's.

Taxable Year

An accounting year is selected by a taxpayer by the filing of his first return and, once chosen, must be retained unless a change is authorized by the Commissioner. If a taxpayer does not elect a fiscal year, he must report on a calendar-year basis. A fiscal year is one which ends on the last day of a month other than December. However, a fiscal year or a calendar year might end on any day of the month by the death of the taxpayer or the termination of a partnership, or the termination of a corporation. See § 6.2.

Inventory Methods

Inventory methods used by accrual-basis farmers are (1) cost; (2) cost or market, whichever is lower; (3) farm-price method; and (4) unit-livestock-price method.

All accrual-basis taxpayers may value inventories at cost or on the basis of cost or market whichever is lower, but farmers have options not available to other taxpayers to use the "farm-price" method, and a method applying only to livestock called the "unit-livestock-price" method. An inventory valuation method once adopted must be followed consistently and may not be changed without the consent of the Commissioner of Internal Revenue, except that a farmer may change from cost, or cost or market to the unit-livestock-price method without consent. See § 6.3.

§ 2.7-2 Year-End Tax Decisions

Cash basis farmers are in a good position to shift income and expenses between taxable years for the purpose of managing taxes. The familiar management procedure is to shift income into the next taxable year and to accelerate expenses and deductions in the current taxable year, but a good tax planner may want to add income and postpone deductions. At all times, a farmer should be constantly on the alert for available deductions.

Tax management is a process of continually surveying the year's income and expense picture. As the tax year nears its end, the proper shifts can be made.

Shifting Income

Deferred sales contracts are an excellent way to defer income to next year. A firm contract to sell grain in the current year can make payment available only in the following year. But if there is a right to reach the income or obtain payment in any way, the income will be taxable in this year. An advance payment on a contract is normally income when received. See § 3.29.

Installment sales of land, machinery and other property can spread any gain over the years of payment. An outright sale resulting in a capital gain would include one-half the gain in this year's income. By spreading the gain over a period of years, it could well be taxed at lower rates, resulting in a lower total tax. See § 3.26.

Commodity credit loans can be treated as income in the year of the loan. Ordinarily, a loan is not income, but a farmer may elect to treat the loan as income when received rather than in the year the crop is finally sold. See § 3.8.

If an election has been made to treat loans as income, a loan is so treated even though the farmer redeems his loan and takes back the grain in the same year. The full amount of the loan (not the net outstanding at the end of the year) is income according to the government. One appeal court held for the government and one for the taxpayer. If the loan is reported as income, the grain obtains a new basis in the amount of the income reported.

Informal income shifting does not work. For example, a farmer had a right to payment for his crop when the buyer inspected the crop on December 31 or January 5, whichever date the farmer selected. He selected January 5—but he was taxed on the income in the year ending December 31 because he could have had the money on that date.

Crop Insurance

A provision which gives relief from doubled income permits a farmer whose crop was destroyed to include the insurance proceeds in the income of the following year if he can prove that he would normally have sold the crop in the following year. This same provision applies to government payments made for crop losses or crop failures. It also applies to livestock sold solely on account of drought but only to the number in excess of normal sales when there was no drought. See § 3.10.

Year-End Deductions

Depreciation deductions are excellent means of reducing taxes. Year-end purchases of depreciable personal property will provide minimum regular depreciation deductions but will give, in most cases, 20% additional first-year depreciation. On a $4000 tractor, this is an $800 deduction. See § 5.2.

Expenses for soil and water conservation, land clearing and fertilizer are controllable year-end expenditures. The full cost of fertilizer and lime, including cost of application, may be deducted as a business expense. Soil and water conservation, terracing, damming, etc. may be written off as current business expenses up to 25% of gross income from farming. See §§ 4.7, 4.13, 4.14.

Land-clearing expenses apply only to new land. This provision fills the gap not covered by soil and water conservation expenses. Soil and water expenses must be made on land being used in farming. Land-clearing expenses are those made on land not previously farmed. However, there is a severe limitation; the expenses can be deducted only up to the lower of $5000 or 25% of the net income from farming. See § 4.14.

See §§ 5.12 and 5.13 on the potential recapture of these deductions as ordinary income against the gain on the sale of the land.

Many items can be paid for at the end of one year or the beginning of the next. The veterinarian, lawyer, hardware store owner, and everyone else understands this.

There are many items that can be purchased near the end of the year. Year-end purchases of supplies, fertilizer, small tools and other routine needs are rarely questioned. Prepaid interest (except "points" on home mortgages) normally will not be deductible. See § 4.5.

Payment of very large sums at the end of the year does draw audit challenges. One of the areas of continuous controversy has been prepayment of feed costs. Several case decisions have permitted a farmer or rancher to make large year-end purchases and deduct these amounts in the year of purchase. However, the courts have laid down several guidelines to determine whether or not the large purchases will hold up. The Service also has stated its position. There are differences between the cases and the Service's ruling. See § 4.6 and the material relating to advance purchases in § 2.2-2.

Investment Credit

The investment credit can be a major year-end tax reducer. Property subject to the investment credit can be bought before or after the turn of the year, thus directing the credit to this year or next.

chapter 3

MANAGING INCOME OF THE FARM BUSINESS

Section

§ 3.1 INTRODUCTION

This chapter deals with the special problems of reporting income. A farmer may have income from:

(1) Farm operations—

(a) Sales of stock in trade (livestock, grain, and other products of farming). Income from these sales ordinarily will be entered on the Schedule F of Form 1040 (discussed in § 3.2).

(b) Custom work, breeding fees, rents, government benefits, patronage dividends, Commodity Credit Corporation loans, and other business income. These items will also be reported on Schedule F (discussed in §§ 3.3 to 3.11).

(2) Sales or other dispositions of business assets—

(a) Some of these assets will qualify for special capital gain treatment (land, buildings, machinery, equipment, livestock held for draft, breeding, dairy or sporting purposes, etc.). Income from the sales of these assets will be reported on Schedule D and Form 4797 (discussed at §§ 3.12 to 3.18).

(b) Other assets used in the farm business will not meet the special requirements for capital gain treatment (draft, breeding, dairy and sporting animals—cattle and horses held for less than 24 months, other livestock held for less than twelve months—and other assets used in the business held for less than the required period. Income from such sales will be reported on Form 4797 under "Ordinary Gains and Losses" (discussed at § 3.12).

(3) Sales or other dispositions of investment or personal assets—

(a) A farmer often has gain from the sale of nonfarm assets such as the farm home, stocks and bonds, commodity futures, personal automobile, household furniture and similar items. Income from the sale of these items will be reported on Schedule D (discussed at §§ 3.20 to 3.22).

(4) Noncash transactions—

(a) There are many noncash transactions in which a farmer might engage, some of which will result in income to him usually measured by the fair market value of the property or services received (discussed in §§ 3.3, 3.27).

(5) Condemnation, involuntary conversions, installment sales, sale of residence or exchanges of property—

(a) In many of these cases, the income can be postponed or spread over a period of years instead of reporting the income in the year of the transaction (discussed in §§ 3.23 to 3.28).

§ 3.2 SALES OF COMMODITIES OR PRODUCTS

Sales of livestock and produce should be segregated into (1) sales of raised items; (2) sales of purchased items; (3) sales of livestock sold because of disease or drought; and (4) sales of livestock used in the business. Gains and losses from sale of business assets should not be reported on the Schedule F (Form 1040).

Schedule F (Form 1040) provides separate schedules for reporting items raised by the farmer and items purchased for resale. The schedule is designed solely to report sales of stock in trade and goods held for sale. An item that is subject to special treatment, such as capital gains or tax-free replacement, should not be included on Schedule F.

Only livestock and produce raised for sale, the costs of which have been deducted, will be reported as income in full. An accrual-basis taxpayer will show his sales of inventoried items in the schedule for accrual-basis taxpayers.

If a purchased item is reported in the raised schedule, the cost of the item, which is an offset against income, will be lost unless specially deducted.

Sales of purchased items should be reported in the purchased schedule where the cost or basis may be claimed. Sales of chickens and other poultry bought when young may be listed in the raised schedule if the cost is (or has been) deducted as an expense.

Feeder animals are likely to be the only sales regularly listed in the schedule for sales of items purchased for resale.

The cost of livestock or any other item purchased for sale is deducted only in the year of sale and the deduction is obtained on the schedule for items purchased for resale. Cash-basis farmers are likely to overlook costs of items purchased several years back.

Payment Before Delivery

Even if the price is not fixed, advance payments for crops or livestock sold should be included in gross income in the year received by cash-basis taxpayers. Any additional amounts are included when received by a cash-basis farmer. The general rule that an accrual-basis taxpayer must report income when the right to the price is fixed does not apply to some sales of goods held primarily for sale. If for all financial reporting purposes the taxpayer chooses to use a method which defers income until some later point, such as delivery, he may use that method for income tax purposes. Thus, income need not be recognized when the contract is signed, nor even when an advance payment is received. Rather, it may be deferred until completion of the contract. There is an exception to that rule which probably will not affect many farmers. All accruals or advance payments related to inventory goods must be included in income by the end of the second taxable year following the year of accrual or receipt. In this last case, the estimated expenses necessary to perform the contract must be deducted when the income is recognized. These rules may be useful to a farmer on the accrual method who wants to receive some cash this year against delivery of his crop next year. However, if this method differs from prior reporting practices, the Commissioner's consent to the change of accounting method must be obtained.

If the transaction is really a loan and not a sale, the amounts received are not included in income until the sale is made, at which time all of the advance payments and any additional payments are income to a cash-basis taxpayer. An accrual-basis taxpayer reports the full sale price when his right to it accrues.

Livestock

Livestock sold or destroyed because of disease should not be reported on Schedule F if there is a gain and the proceeds are to be reinvested in similar animals. This is treated as an involuntary conversion and the animals may be replaced without recognition of gain. See § 3.24. Even if the animals are not replaced, recognition of gain on sales due solely to drought may be deferred for one year to the extent that drought sales exceed normal sales.

Animals held for draft, breeding, dairy, or sporting purposes, whether held more or less than 24 months (cattle and horses) or twelve months (other livestock), are properly reported on Form 4797. Even if the animal was held for less than twelve or 24 months and was sold at a loss, it should not be reported on the Schedule F because an adjustment for self-employment income will be necessary.

Whether raised or purchased, inventoried or not, draft, breeding, dairy, and sporting animals held for the full period must be included with sales of assets used in the business on Form 4797. If included in the sales of goods held for sale, capital gain benefit will be lost. Gains or losses on animals held less than the required twelve or 24 months are reported on Form 4797 under "Ordinary Gains and Losses." See § 3.14.

Assets Used in the Business

Sales of land and property of a character subject to depreciation held for the long-term capital gain period are not reported on Schedule F. They too, are entitled to special treatment and may also involve a recapture of depreciation or land-expense deductions if sold at a gain. These sales are reported only on Form 4797. Even if these assets were held for less than the long-term period, they are to be reported on Form 4797 under "Ordinary Gains and Losses." See § 3.12.

Before the 1976 Act, the assets needed to be held more than six months. However, under the new Act, the holding period is extended to more than nine months in 1977 and more than one year in 1978 and later years.

No asset subject to depreciation should be listed on the Schedule F.

Farm Income Tax Manual, §§ 301, 714.

§ 3.3 MERCHANDISE RECEIVED

Items of merchandise received by a farmer will ordinarily produce income to the extent of their fair market value.

If farm produce is traded or exchanged or sold for property, the fair market value of the merchandise received is treated the same as cash income. If the merchandise received is then used in the farm business, the value of it is deducted just as if it were a cash purchase.

If a farmer received services and paid for them in produce, the value of the service is income and, if the cost of the service would be deductible, the value of the service is deductible. For example, if he pays for his bookkeeping work with produce, he has income to report and a corresponding deduction. If the service received is not business related, there is income but no corresponding deduction.

The same principle applies to merchandise received in exchange for services rendered by the farmer. The amount of income to be reported is the fair market value of the merchandise received. There is, of course, no deduction for the value of the farmer's own labor, but any cost of earning the income is deductible.

Government payments received in kind are income. See § 3.6.

Farm Income Tax Manual, § 303.

§ 3.4 RENTS RECEIVED IN CROP SHARES

Crop-share rents are reported as income when they are fed to livestock, sold or otherwise reduced to cash or its equivalent.

Neither cash- nor accrual-basis farmers report rent received in crop shares until the year in which the crop shares are reduced to cash or its equivalent. Accrual-basis taxpayers do not inventory crop-share rents on hand at the end of the year. Since 1975, the Internal Revenue Service has treated a gift of crop shares to charity as a realization of income. This position was first asserted and then abandoned in the 1950's after the Service lost some cases. The revived position may not survive litigation any better than it did in the 1950's, but in the meantime it reflects the Service's view.

Because crop-share rents are not income until sold and are not included in inventory, a cash- or accrual-basis farmer may determine when he shall report the income because he has control over the time of reduction of the crop rents to money. "Crop share" means share of crops or livestock.

Before 1975, if crops were fed to livestock, the income was realized in the gain on the sale of the animals. Since 1975, however, the Service argues that gain occurs on the feeding and that there is an offsetting deduction. A cash-basis taxpayer would have a wash, but an accrual-basis taxpayer would include the value of the animal (if not disposed of) in inventory at the end of the year.

Death of Farm Landlord

When crop-share rents pass at the death of a farm landlord who did not materially participate in farm production (§ 7.5) and the estate (or other successor to the decedent) sells the crops, it must report the income that the decedent would have reported had he lived to sell the crops. The estate or successor is entitled to a deduction for some or all of the amount of the death taxes paid on the value of the crops.

If the landlord dies during a rent period, the value of his interest in crops and livestock at the date of death is included in his gross estate for federal estate tax purposes as accrued rent. When such crops and livestock are sold after his death, the portion of the proceeds allocable to the period before death is treated as income earned before death. The remainder of the proceeds is ordinary income earned by the estate or successor in interest as the new owner. The estate or successor, of course, may deduct any ordinary expenses incurred in raising or selling the crops or livestock. The practical difference is that the estate or successor is entitled to an income tax deduction for the amount of federal, and, where death occurs after 1976, state death taxes paid on the income "earned" before death, but not on income earned after death which was not subject to estate tax.

Crop Rentals Fed

Under the 1975 ruling, if crops received as rent are fed to livestock, the transaction should be treated as a "wash" for income tax purposes. If the value of the crop rents fed is regarded as income, there is an offsetting feed deduction in the same amount. If this ruling does not stand up and the crops were received by the landlord before death and fed to his own livestock (not rental share animals), there will be no income or deduction. In either case, where death occurs after 1976 the livestock should take the decedent's basis. In computing gain, this amount may be adjusted for increases in value before January 1, 1977, and for state and federal death taxes attributable to increases in value after that date.

If the crops are fed before the death of the landlord to animals which were rental shares, and if the 1975 ruling does not stand up, the crop rents disappear but the livestock would be "rents" having the quality of income. The rental income would be reported by the estate or successor on subsequent sale of the animals. If the estate fed the crop rents to livestock (whether rental shares or not), no income should be reported and no feed deduction should be claimed because there would be no basis for the crops. The income inhering in the crops would be realized on the sale of the livestock. If the livestock acquired by the estate were not received by the decedent as rents, the only pre-death income on their sale would be the amount of crop-share rents fed. The costs of raising after the death of the landlord (except for the crop rents fed) would be deductible and there would be an additional deduction against the proceeds of the sale for the federal and, where death occurs after 1976, state death taxes paid on the crop rents used as feed.

If the 1975 ruling is correct, the feeding itself results in income and a corresponding deduction. If the feeding occurred before death to animals which were themselves crop shares, there would not seem to be any deduction for death taxes paid on the value represented by the feed because it would no longer be income in respect of a decedent. This result seems more clearly to follow if the animals are not also crop shares.

The same results should follow even though the farm landlord is on the accrual basis.

Material Participation Lease

Presumably, neither the old law or the 1975 ruling applies to crops and livestock received under a "material participation" lease. Therefore, the estate of a "materially participating" landlord would, after 1976, have a "carry-over basis" equal to the decedent's basis before death. For computing gain, the basis may be increased by increases in value before January 1, 1977, and for federal and state death taxes on increases in value after that date. If the landlord died before 1977, the estate would have a basis equal to the fair market value at death. See § 5.8.

Gift of Crop-Share Rents

While the law surrounding gifts of crop shares is not entirely clear, the "material participation" lease appears to be an important concept. Under such a lease, the landlord seems to take on the characteristics of a farmer. If the landlord is a farmer, his crop shares are treated as crops in his hands, and generally speaking he may assign them to either a charity or a family member without being subjected to tax on the value of the crops either at the time of the gift or when the crops are sold.

On the other hand, if the lease does not require "material participation," the crop shares are deemed to be "rent." A transfer of "rent" usually will result in a tax to the landlord, either at the time of assignment (under the 1975 ruling) or when the product is sold (under prior law). These rules are not hard and fast, but the gift to charity should receive treatment no less favorable than a gift to a family member.

Generally, the rule is that the landlord who transfers **crop rents** to a charity will report income and a charitable deduction in the same amounts—in effect, a wash. The cash-basis farmer or materially participating landlord who gives **crops** will have no charitable deduction because he has no basis in the crops. See § 5.8.

Farm Income Tax Manual, § 305.

§ 3.5 WOOD AND LUMBER

Sales of wood and lumber are treated as sales of other farm produce except that cost may be recovered through depletion and in some cases the value of the timber growth may be entitled to capital gain treatment.

Sales of wood or lumber products severed from the soil produce ordinary income subject to a deduction for depletion. See § 5.14.

Proceeds of the sale of tree products (e.g., fruits, nuts, syrup) are treated as ordinary income. A small, handy reference is the U.S.D.A. pamphlet "The Timber Owner and His Federal Income Tax," Agriculture Handbook No. 274.

If the wood or lumber sold was purchased or if the farmer has made an election to treat the cutting of the timber as capital gain, the sale should be reported as a sale of a purchased item and the basis is the cost of the wood or lumber purchased or the fair market value of the timber at the time of cutting. Added to the latter figure will be any additional cost of preparing the lumber for sale unless those amounts had been deducted as expenses. See §§ 3.17, 3.18.

Where the farmer sells the standing trees with the land or disposes of the standing timber under a cutting contract and elects to treat the royalty payments as capital gain, the income is properly reported on Form 4797, not on the farm schedule. See §§ 3.17, 3.18.

Farm Income Tax Manual, § 306.

§ 3.6 AGRICULTURAL PROGRAM PAYMENTS

Payments received under government programs are income.

Government payments are income whether received in cash or in materials. The value of fertilizer or lime is income, but there is an offsetting deduction. Payments based on improvements such as concrete outlets or tile are income, but the cost is recovered by depreciation.

Conservation reserve program payments are income, including payments to cover the cost of establishing conservation practices, the annual rental on acreage placed in the conservation reserve and the fair market value of seedling trees, shrubs, and seeds received. Negotiable certificates received are includable at their face value in the year of receipt. If certificates are used to discharge a commodity credit loan in excess of their face amount, the excess is income in the year the loan is discharged.

The cash equivalent of certificates for wheat program payments must be included in income when received or when available, whichever is earlier.

Farm Income Tax Manual, § 307.

§ 3.7 PATRONAGE DIVIDENDS

Dividends and rebates received by a patron are income but there are some special rules to determine when the income is reported.

A cash dividend received on products sold amounts to an additional price received for a product of the farm and is fully taxable. The patron reports as current income the cash, the stated amount of a "qualified" written notice of allocation, and the fair market value of other property (except a nonqualified written notice of allocation) received as a patronage dividend (or a nonpatronage distribution from a tax-exempt co-operative).

When the dividend is, in effect, a rebate on the cost of something used by the farmer in his business, the purchased article was either (1) a current expense, or (2) a capital investment. If it was deducted as a current expense, the dividend is income.

If the dividend or rebate relates to a capital asset or property used in the trade or business, the cost of the property is reduced by the dividend. The adjustment is made as of the first day of the taxable year in which the dividend is received. A rebate on depreciable property will reduce the present adjusted basis of the property. The rebate will, in effect, be spread over a period of time since the reduction of the depreciation deduction increases ordinary income over the years.

If the dividend exceeds the unrecovered cost or the patron no longer owns the property, the rebate is treated as income.

A dividend paid by a purchasing association on consumer goods, or on supplies, equipment or services (the cost of which was not deductible) is not taxable.

Dividend Allocations

Written notices of allocation may be capital stock, revolving fund certificates, retain certificates, certificates of indebtedness, letters of advice or other written notices which disclose to the recipients the dollar amount allocated and the portion which constitutes a patronage dividend (or a nonpatronage distribution by a tax-exempt co-operative).

The notice is "qualified" if (1) 20% of the total dividend is paid in cash or by qualified check, and (2) it is redeemable in cash within ninety days of issue or the patron has consented to report it as income.

If a written notice of allocation is not "qualified," it is not income when issued. Income is reported when the item is redeemed, sold, or disposed of.

When a taxable allocation is redeemed for less than its face value, the patron has an ordinary loss if his membership in the co-operative was for business purposes.

Per unit retain certificates are also treated as "qualified" and "nonqualified." If the patron has consented to take the value of the certificate into income, it is qualified; others are nonqualified. The qualified certificate is included in income in the year received. The nonqualified certificate is included in income when it is disposed of or redeemed. The co-operative reduces current income by the amount of qualified certificates but may deduct the nonqualified only when redeemed.

A qualified notice dividend is included in income even though the amount of the dividend is applied to the price of items bought by the patron from the co-operative or applied to his membership dues.

Cancellation of outstanding marketing credits of a patron is an ordinary loss.

Farm Income Tax Manual, § 308.

§ 3.8 COMMODITY CREDIT LOANS

A loan received from the Commodity Credit Corporation is not income unless the farmer makes a special election.

Loans received from the Commodity Credit Corporation may be treated as income in the year received if the farmer so elects. Unless such election is made, the receipt of the loan is not income and he claims no deduction when he repays the loan.

The farmer who has made an election to treat CCC loans as income reports the income in the taxable year in which he actually receives the proceeds of the loan. If he has not made the election, he has no taxable income until the year in which the commodity is sold.

When the crop is sold by a farmer who made an election to treat the loan as income, he reports additional income only to the extent that the sale price exceeds the amount of the loan. If the sale price is less than the loan, there is no deductible tax loss unless the farmer is required to, and does, satisfy the deficiency during the taxable year.

Only after a farmer has treated a loan as income is he bound to continue the pattern. A farmer who has always treated the loans as loans may make the election to treat the loans as income at any time. It is made by treating the loan as income on an income tax return filed on time and attaching a statement so electing.

If the farmer has elected to include CCC loans as income, the face amount of a "certificate of interest" issued by the Corporation is not a loan and is not income until the amounts are actually paid.

Loan Redeemed

If a farmer (who has elected to treat loans as income) receives a loan and during the same year redeems part of the commodity, the government says that the reportable income is the full amount of the loan, not the net loan outstanding at the end of the year. The crop then has a basis equal to the income reported. One case supports the government and one is contrary.

Fees paid for storage by the Commodity Credit Corporation are self-employment income to a farm operator or a landlord who materially participates in the farm operations but are rental income to a landlord who does not materially participate.

If a farm operator stores grain for other farmers, the fees received from the storage are regarded as income from the business of warehousing which is a separate and distinct business from that of farming. But when he stores his own grain the fees are regarded as farm income.

Farm Income Tax Manual, § 309.

§ 3.9 RECOVERIES OF DEDUCTED ITEMS

The recovery of an item which was deducted in a previous year is treated as income to the extent that the deduction gave the farmer a tax benefit in the prior year.

As a general rule, a recovery of an item previously deducted is to be treated as income. For example, if a deduction was properly claimed in 1975 and the taxpayer received a refund of part of it in 1976 he would report the refund as income in 1976. However, he reports as income only so much of the recovery as, when previously deducted, caused a decrease in tax for the prior year. If the farmer's other deductions exceeded his income in 1975, so that this deduction did him no good, none of the recovery will be income. This is the so-called "tax benefit" rule.

If the recovery or refund is received in the same year that the deduction would be claimed, generally the deduction is reduced by the recovery. In most cases, it would not matter if the deduction were claimed in full and the recovery reported as income, which may be the simplest procedure.

Refunds of federal income taxes are not income for federal tax purposes because the taxes were not deducted. The interest received on such refunds is ordinary income. Further, if interest was paid on federal taxes due and the taxes plus interest were later refunded, the interest paid was deductible, so the interest recovered is income subject to the "tax benefit" rule.

See § 6.6 for recovery of gasoline tax as an income tax credit.

Farm Income Tax Manual, § 310.

§ 3.10 INSURANCE PROCEEDS

The proceeds of fire or casualty insurance are treated in the same manner as the proceeds of a sale and will produce ordinary income or loss or capital gain or loss depending on the assets. Some gains may be postponed if the asset is replaced.

Proceeds of insurance are generally treated as payment for the asset destroyed or damaged, and the gain or loss is measured by the difference between the proceeds and the adjusted basis of the property. See § 4.17. For example, all of the proceeds of hail or fire insurance on growing crops which have no basis is ordinary income. Insurance received on livestock held for sale is treated as the proceeds of a sale.

Crop Insurance

The proceeds of insurance on the destruction of a crop may be reported in the following year if that is the year the crop would normally have been sold. The 1976 Act allows crop disaster payments to be included as insurance proceeds. Beginning with 1976, the same rule may be applied to the proceeds of livestock sold on account of drought to the extent that the drought sales exceed normal sales.

Regulations provide that the election must be made at the time of filing the return for the year of destruction or damage by attaching a separate statement signed by the taxpayer to the return. The statement must contain the following information: (1) a declaration that the taxpayer is making an election under section 451(d); (2) the specific crop or crops destroyed or damaged; (3) a declaration that under the taxpayer's normal practice the income from the crops which were destroyed or damaged would have been included in his gross income for a taxable year following the year of such destruction or damage; (4) the cause of the destruction or damage to the crops and date occurred; (5) the total amount of payment received from the insurance companies with respect to each crop and the date each payment was received; and (6) the names of the insurance companies which made the payments.

New regulations will apply similar rules to crop disaster payments and livestock sold because of drought.

Insurance for Living Expenses

Prior to 1969, insurance paid to cover the living expenses of the taxpayer and his family when the residence was damaged or destroyed by casualty were treated as income. Thereafter Section 123 of the Code excludes from income the excess of the actual expenses over the normal living expenses that would have been incurred.

Casualty Losses

If insurance proceeds are received as a result of the destruction of a capital asset, or a Section 1231 asset used in the business, held for the long-term capital gain period, the gain or loss on the "involuntary conversion" is included in a special aggregation of casualty gains and losses. If the aggregate of the casualty and theft gains and losses results in a gain, all of the transactions are included in the Section 1231 aggregation. If the aggregate of the casualty and theft transactions result in a net loss, the transactions are not included under Section 1231, and the net loss is fully deductible. See § 3.12.

Before 1977, the asset had to be held more than six months; in 1977, more than nine months; thereafter, more than one year.

Avoiding Gain

The proceeds of insurance received on the destruction of any asset may be reinvested in similar property free of tax. See § 3.24.

If there is a partial destruction, the insurance proceeds reduce the basis of the asset. Any excess of proceeds over the adjusted basis would be gain, either ordinary or capital, depending upon the nature of the asset. Proceeds used for rebuilding may be treated as reinvested free of tax in similar property.

If the insurance is less than the basis of property that is totally destroyed, the taxpayer has suffered a loss. Whether the loss is ordinary or capital or falls under Section 1231 depends on the nature of the asset and whether casualties and thefts produced a net gain or net loss. See §§ 3.12, 3.20.

Farm Income Tax Manual, § 311.

§ 3.11 PRODUCTS DONATED

A gift of farm produce to a charity does not result in income to the donor, but the charitable deduction is limited to the basis of the property.

A gift of farm produce to an individual does not result in income to the donor, but he may lose some of his deductions for the cost of raising the produce. A private gift of crop-share rents will not divert income from the donor.

A gift of appreciated property (which would produce ordinary income on sale) to a charity will be deductible only to the extent of the basis. If a cash-basis farmer donates raised wheat, his basis and deduction will be zero. Even after the ruling on crop shares discussed at § 3.4, the Service is not expected to argue that there is income to the donor on the transfer of such income property. Thus, the donation should not result in income or deduction to a cash-basis taxpayer.

A cash-basis taxpayer does not adjust his deductible expenses for costs incurred in producing the item. An accrual-basis taxpayer must remove the item from his inventory and eliminate from cost of goods sold any costs of producing the item.

If the property given to a charity would produce long-term capital gain on sale, no income arises from the gift. If the gift is tangible personal property which the charity does not use in its exempt activity, the deduction is the value of the property reduced by 50% of the potential long-term gain. Otherwise, the full deduction is allowed on gifts to public charities, and the deduction is reduced by half of the potential capital gain on gifts to some private foundations. The deduction is always reduced by the potential recaptures of depreciation, land expense, and EDA. See §§ 5.9 to 5.13.

Private Gifts

In the case of private gifts of property held for sale, any item with respect to the donated product which has been deducted as a business cost must be eliminated. If the item is inventory property, the beginning inventory must be reduced to remove the asset and the costs pertaining to it. The donee's basis for the asset for computing gain to the donee is the donor's basis increased by the donor's costs which were not deductible; and, for computing loss to the donee, the lower of the donor's basis or the fair market value at the time of the gift. Some of the gift taxes paid on the gift may be added to basis, but only to the extent that such addition does not raise the basis above fair market value. The basis for depreciation, if any, is the basis for computing gain. See § 5.8. Depreciation on a business asset and land deductions which would have been recaptured if the donor sold the property will be recaptured from the donee when he sells it. See §§ 5.10 to 5.12. A special rule for recapture of a donor's "excess deductions account" is discussed at § 5.13.

Crop-Share Rents

See the discussion at § 3.4.

Gift of Remainder Interest

The transfer of a remainder interest in property to a charity where the income interest is retained by the donor or given to a member of his family is generally subject to stringent rules on deductions for gifts of remainder interests. However, the remainder interest in a farm or a residence is not subject to these rules and may still be transferred to charity and a deduction taken.

Farm Income Tax Manual, § 312.

§ 3.12 CAPITAL GAINS ON BUSINESS ASSETS

Capital gains and losses arise from transactions in two separate types of property. A sale or exchange of a capital asset generally results in a capital gain or loss.

"Long-term" capital gains receive favored tax treatment. Capital losses may not be fully deductible.

Gains and losses from sales and other dispositions of assets used in the business of farming (such as land and breeding livestock) and involuntary conversions of capital assets are aggregated together. If the aggregate is a gain it will be a capital gain. If the aggregate of these transactions is a loss, it will usually be an ordinary loss.

For convenience, these latter types of assets or transactions are referred to as "Section 1231 assets" or "Section 1231 transactions" from the section of the Internal Revenue Code which provides the special method of taxing.

Capital assets are defined in negative, rather than affirmative terms. In substance the law describes capital assets as: ALL PROPERTY held by the taxpayer (whether or not in connection with his trade or business) except:

(1) Stock in trade or other property properly included in inventory, and property held primarily for sale to customers in the ordinary course of the taxpayer's trade or business;

(2) Property used in the trade or business, of a character subject to depreciation, and real property used in the trade or business (Section 1231 items); and

(3) When held by the creator, a copyright, a literary, musical or artistic composition or similar property, letters and memoranda, and certain accounts or notes receivable, and federal, state and municipal obligations issued on a discount basis and payable without interest at a fixed maturity date not exceeding one year from the date of issue.

The first exception excludes most farm produce which gives rise to ordinary income. The second exception excludes farm equipment, land, buildings and depreciable livestock covered by Section 1231.

Gains from transactions involving capital assets held for more than a minimum period receive specially favored tax treatment. Before 1977, the minimum period was six months; now it is nine months, and in 1978 and later years it will be one year. In general, only half of the gain is subject to tax, and in no event may the tax rate on the full gain exceed 25% of the first $50,000 of gain in any year and 35%

on the excess. The untaxed one-half of capital gains may, however, be subject to the minimum tax on tax preferences.

When a farmer speaks of "capital gains," he is referring to this special tax treatment. In technical tax language, a gain from the sale of a capital asset held for more than the minimum period is a "long-term gain."

Gains from capital assets held for the minimum period or less are called "short-term gains." The full amount of the short-term gain is included in income subject to the regular rates on ordinary income.

Losses arising from transactions involving capital assets are subject to some limitations on deductibility. Capital losses offset current capital gains (both long-term and short-term). Any excess capital losses may offset other income up to $1000 in 1976, $2000 in 1977 and $3000 in 1978 and later years. Unused capital losses may be carried forward to offset capital gains and these amounts of ordinary income in future years. Short-term capital losses offset ordinary income dollar for dollar but long-term losses are reduced by half, so that a dollar of long-term loss will offset only fifty cents of ordinary income.

Section 1231 Transactions

The gains and losses from transactions described in Section 1231 are aggregated together to determine whether there is a net gain or a net loss.

Generally, the Section 1231 transactions are sales, exchanges and involuntary conversions of draft, breeding, dairy, and sporting cattle or horses owned for 24 months, other livestock owned for twelve months, and land and other assets used in the farm business held for more than six months in 1976, nine months in 1977, and one year in later years. Section 1231 also includes (1) involuntary conversions of capital assets held for more than the minimum period, (2) timber sales under royalty contracts, (3) the sale of growing crops with the land, and (4) if the taxpayer elects, the cutting of timber treated as a sale. See § 3.13 for details.

If the result of aggregating all Section 1231 transactions is a net gain, all of the transactions are treated as capital transactions and the net gain is a long-term capital gain. If the net result of the aggregation of all of the Section 1231 transactions is a loss, then each asset is treated as producing ordinary gain or loss. As a practical matter, a net loss from the aggregation of Section 1231 transactions will be fully deductible.

A preliminary step must be made. First, all gains and losses from casualty or theft are aggregated. If the result of this initial aggregation is a net gain, all such transactions are included in Section 1231. However, if the initial aggregation is a net loss, then none of the transactions arising from casualty or theft are included in the Section 1231 mix. In effect, a net loss from casualties and thefts results in full deduction rather than merely offsetting other Section 1231 gains.

Gains and losses which are not recognized do not fall under Section 1231. For example, if a farm is exchanged for another farm the gain is not recognized, and it is not included in the Section 1231 computations.

It is to the obvious advantage of the farmer to try to realize his Section 1231 gains and losses in different years so that gains will be capital gains and losses will be deductible in full.

Involuntary Conversions

If there is an involuntary conversion, for example, when insurance is collected because property is destroyed or stolen, or when the government takes property by condemnation, Section 1231 (subject to the casualty loss modification) applies whether the property is business property or a capital asset. This is the only instance where a "capital asset" is put in the same class with transactions involving assets "used in the business." If the farmer has no gains from Section 1231 transactions, an allowable loss on condemnation, sale under threat of condemnation, or destruction or theft of a capital asset will result in an ordinary, fully deductible loss. See §§ 3.24, 3.25.

Computing Gain or Loss

Gain or loss when property is sold, exchanged, or otherwise disposed of is the difference between the cost (or other basis of the property) and the amount of money and the fair market value of other property received.

The basis on which gain or loss is computed is figured in exactly the same way as basis of property for depreciation. The discussion on computation of the basis appears in § 5.8, Basis for Depreciation.

When capital assets or Section 1231 assets are sold, special rules apply. These may mean that only part of the gain or loss is taxable or deductible. Losses on some transactions are not deductible at all. See § 4.21. Unless some such special rules apply, the difference between what is received and the basis to the taxpayer is ordinary income fully taxable or ordinary loss fully deductible.

Capital Gains and Losses

The capital gain and loss provisions apply to sales or exchanges (or dispositions defined as a sale or exchange) of all capital assets except those capital assets involuntarily converted which are included under Section 1231. In addition, if the aggregate of the Section 1231 transactions results in a gain, each Section 1231 transaction is treated as a sale or exchange of a capital asset.

Schedule D (Form 1040) and Form 4797 are used to report several types of transactions:

(1) Dispositions of capital assets held six months (nine months in 1977; one year in 1978 and later years) or less,

(2) Sales or exchanges of capital assets held for more than six months (nine months in 1977; one year in later years) including Section 1231 transactions if the aggregate is a gain,

(3) Disposition of assets which are not capital assets; this will include most Section 1231 assets if the aggregate is a loss. It will also include assets on which there is an element of gain to be recaptured as ordinary income.

Gains and losses from sales and exchanges of capital assets held for six months (nine months in 1977; one year in later years) or less are "short-term gains and losses." These are separately computed and reported in the "short-term" schedule on the Form 1040, Schedule D.

Gains and losses from sales and exchanges of capital assets held for more than this minimum period, including Section 1231 transactions, are called "long-term gains and losses." These are separately computed and reported on the "long-term" schedule of Form 1040, Schedule D.

The separate computations produce a net short-term gain or loss and a net long-term gain or loss. If there is a **net short-term gain** in excess of the net long-term loss, the full excess is treated as ordinary income. If there is a **net long-term gain** in excess of the net short-term loss, half of the excess is included in income, or, stated more precisely, 50% of the excess is deducted. In any event, the tax on the full amount of the net long-term gain may not exceed 25% of the first $50,000 ($25,000 for each spouse on a separate return) in any year.

This alternative tax on long-term capital gains does not apply to gains in excess of $50,000, but the practical maximum on such gains is 35 per cent. The maximum tax rate on ordinary income is 70%, and half of the long-term capital gain will be subject to ordinary income tax rates. The 25% maximum still applies to the first $50,000 of long-term capital gains in any year. The untaxed one-half of the gain is a tax preference and may be subject to the minimum tax.

Capital Loss Carry-Over

If the aggregate of the long- and short-term gains and losses results in a loss, the loss is deductible up to (a) before 1977, $1000, (b) in 1977, $2000, and (c) in later years, $3000 (or taxable income if less than these amounts). If this aggregate is short-term losses, they offset ordinary income dollar for dollar. If this aggregate is long-term losses only one-half may be used against these amounts of ordinary income; in effect, $1 of long-term loss offsets fifty cents of ordinary income.

Any excess loss may be carried forward to future years and used to offset capital gains of those years plus $1000 before 1977, $2000 in 1977, and $3000 thereafter (or taxable income if less than these amounts). In carrying capital losses forward to future years, short-term and long-term losses retain their separate character. Short-term losses are applied in the carry-over year to offset, first, short-term gains, then long-term gains, then ordinary income. Similarly a long-term loss is applied to long-term gain, then short-term gain and then to ordinary income. Of course, the limitation of the offset against ordinary income also applies on the carry-over; $1 of long-term loss offsets fifty cents of ordinary income. Short-term losses are applied before long-term losses.

The first schedules of the Form 1040, Schedule D, do this job of aggregation of the short- and long-term gains and losses. The final figure will be a capital loss or it will be a capital gain which will be reduced by 50% of the long-term capital gain. The alternative tax computation on Schedule D ensures that long-term capital gains of $50,000 or less will not be taxed in excess of 25% of the gain.

No Capital Gains for Related Taxpayers

There is no long-term capital gain treatment on sales of depreciable assets between spouses or between an individual and his controlled corporation or between two controlled corporations; the gain is ordinary income notwithstanding the capital asset or Section 1231 rules. Ordinary income also results from sales between a partner and a partnership in which he has an 80% interest or between two partnerships in which the same persons have an 80% interest. Losses on sales between related taxpayers may also be disallowed. See § 4.21.

Farm Income Tax Manual, §§ 315, 333.

§ 3.13 PROPERTY USED IN THE FARM BUSINESS

Section 1231 of the Internal Revenue Code is a special provision intended to extend tax relief to gains and losses (1) from the disposition of property used in the trade or business, (2) under some circumstances, from the disposition of livestock, growing crops, timber, coal, and iron ore, and (3) from the involuntary conversion of capital assets.

In general, the gains and losses from property sold, exchanged, or involuntarily converted receive the special tax treatment of Section 1231 if the property was:

(1) Land, or property of a character subject to depreciation,

(2) Used in the trade or business,

(3) Held for more than six months before 1977, nine months in 1977 and one year in 1978 and later years,

(4) Not properly includible in inventory,

(5) Not held primarily for sale in the ordinary course of business, and

(6) Not a copyright or a literary, musical or artistic composition, a letter or memorandum or similar property in the hands of the creator of the property or the person for whom it was prepared, or in the hands of a person whose basis is the same as the creator of the property.

Certain other property is specifically declared to be "property used in the trade or business" and therefore eligible for Section 1231 benefits:

(1) Livestock held for draft, dairy, breeding or sporting purposes and held:

    (a) cattle and horses—for more than 24 months from the date of acquisition, and

    (b) other livestock—for more than twelve months from the date of acquisition.

(2) Unharvested crops sold or exchanged (or involuntarily converted) with the land at the same time and to the same person, if the land was held for more than six months before 1977, nine months in 1977, and one year thereafter;

(3) Timber owned for more than six months before 1977, nine months in 1977, and one year thereafter before the beginning of the year in which it is cut, if the owner elects to treat the cutting as a sale;

(4) Timber, owned for more than six months before 1977, nine months in 1977, and one year thereafter sold under a royalty contract;

(5) Coal and iron ore owned for more than six months before 1977, nine months in 1977, and one year thereafter sold under a royalty contract.

Basically, there are just two requirements to be met in qualifying property for Section 1231 treatment—(1) that it be used in the business, and (2) that it be owned more than the minimum period (livestock for twenty-four or twelve months).

Used in the Trade or Business

To be "used in the business" an asset must be used as a part of the farming operation, for example, farm land and buildings, machinery and equipment, or be one of the classes of property specifically declared to be so used, such as timber, coal, livestock, and unharvested crops under certain circumstances. These special types of property are discussed separately in subsequent sections. Generally a business is an activity in which a person regularly engages for profit and from which he seeks all or part of his livelihood.

The property need not be used solely for business use. If an asset is used only partially for business purposes, such as the farmer's automobile, or a power mower adaptable for both farm and lawn use, the asset is divided into two parts according to the percentage of business and personal use. For example, if the car were used 60% for farming business and 40% for personal uses, 60% of the sale price would be treated as the sale of a Section 1231 asset and 40% would be treated as the sale of a capital asset. If there is a loss, only the 60% business portion would be considered. The 40% of the loss attributable to the personal portion would not be deductible because it was not incurred in a trade or business or in the production of income.

A similar result follows with respect to a farm residence in which rooms are used for a business office. The part devoted to business use is a Section 1231 asset and the rest is a capital asset.

Property Subject to Depreciation

Property "of a character subject to depreciation" is a Section 1231 asset. The quoted phrase does not mean that a particular item of property must have been depreciated. Rather, it need only be "of a character" subject to depreciation. For example, raised breeding animals with a zero basis to a cash-basis taxpayer or inventoried breeding animals of an accrual-basis taxpayer are of a character subject to depreciation. Thus, although they have not been depreciated, they qualify as Section 1231 assets.

Similarly, while the cost of a hammer, an anvil, or other small tools or equipment may have been deducted as a current expense, such items normally have a useful life of more than one year and are "of a character" subject to depreciation. Consequently, when the farmer has a sale in which the tools and small equipment accumulated over many years are sold, he will treat the "gain" under Section 1231 on the grounds that these assets "are of a character" subject to depreciation.

Tractors, combines and other machines used by the farmer and in which he has an investment subject to a yearly allowance for depreciation qualify for Section 1231 treatment upon sale, exchange, or involuntary conversion. Even though these assets may have been completely depreciated so that the basis has been reduced to zero, they are still property "of a character subject to depreciation."

Assets Held for Sale

Assets "properly includible in inventory" are excluded from the benefits of Section 1231 in general, but draft, dairy, breeding, or sporting animals are not excluded even when inventoried. Since livestock is specifically defined as property used in the trade or business when held for these purposes, it makes no difference whether they are included in inventory. Their inherent character as subject to depreciation is not changed even for accrual taxpayers who include such animals in inventory.

The purpose of the exclusion of inventory items from the benefit of Section 1231 is to prevent the ordinary products of the farmer's business or any other trade or business from getting the tax bargain treatment. Normally the type of thing which a businessman has on his shelves to sell to customers should produce ordinary income. Similarly, the farmer must exclude from Section 1231 all inventoried or inventory-type products of his business except draft, breeding, dairy and sporting livestock.

The elimination from Section 1231 of assets held primarily for sale in the ordinary course of business applies the same rules to the products of the cash-basis farmer as the phrase "not properly includible in inventory" does to the products of the accrual-basis farmer.

Holding Period

Long-term capital gain results only if the asset was held for more than a minimum period. Before 1977, the period was six months; in 1977, it is nine months; in 1978 and later years, one year. The computation of the minimum holding period (as well as the twelve-months, or the 24-month holding period for livestock) is determined by the use of calendar months, regardless of the actual number of days in the month. April 30 to May 31 is a month; so is January 31 to February 28.

In making the computation, the day of acquisition is excluded and the day of disposition is included. An asset bought on February 15 and sold on November 15 would be held for exactly nine months. It would not have been held for **more** than nine months. If sold on November 16, it would have been held for more than nine months. The intervention of holidays or less than full business days has no effect upon the computation.

See § 5.8 for holding period on carry-over basis assets.

Farm Income Tax Manual, §§ 315 to 325.

§ 3.14 DRAFT, BREEDING, DAIRY, AND SPORTING LIVESTOCK

Livestock, regardless of age, held for draft, breeding, dairy, or sporting purposes and held for 24 months (cattle and horses) or twelve months (other livestock) are assets used in the farm business which qualify for the special capital gains treatment. Poultry, fish, frogs, and reptiles are not "livestock."

Section 1231 in defining "property used in the trade or business" states: "Such term includes—(A) cattle and horses, regardless of age, held by the taxpayer for draft, breeding, dairy, or sporting purposes, and held by him for 24 months or more from the date of acquisition, and (B) other livestock, regardless of age, held by the taxpayer for draft, breeding, dairy or sporting purposes, and held by him for twelve months or more from the date of acquisition. Such term does not include poultry."

The term "other livestock" includes hogs, mules, donkeys, sheep, goats, fur-bearing animals, and other mammals. It does not include poultry, chickens, turkeys, pigeons, geese, other birds, fish, frogs, and reptiles. Pelts from culled breeding animals are entitled to the special treatment.

Criteria for Section 1231 Livestock

Determining whether particular sales of livestock come under Section 1231 depends upon proof that the farmer or rancher actually held the animals for the required purpose. There seems to be no argument about draft animals and little about dairy animals.

The word "sporting" was added to the descriptive phrase of Section 1231 making it read "held by the taxpayer for draft, breeding, dairy, or sporting purposes." A new area of definition may be opened up; presumably "sporting" includes all uses that involve "sport" to a user or spectator. The major purpose of the addition was to cover race and show horses and dogs.

The arguments in the past have been concerned primarily with breeding animals, and particularly cattle. Many of these have disappeared with the extension of the holding period to two years since cattle at that point are very near maturity.

The law does not require actual use of the animal for draft, dairy, breeding, or sporting purposes but merely that the animal be held for those purposes and owned for the statutory period. It has been held that the owner of a dairy and breeding herd, who culled out as rapidly as possible calves not measuring up to standard, could treat these sales under Section 1231 even though the animals had never been bred or used for breeding.

The courts declined to penalize farmers and ranchers who could detect inferior animals before they were bred. It is not necessary for the animals to be used in the dairy or breeding herd. The question is: "For what purpose were the animals held?"

There was a definite shift in the thinking of the courts from a position that an animal could not be a part of the herd until it had proved qualified to a recognition that an animal could become part of a herd (even though not yet used) until proved unqualified. It is possible as an animal matures to make judgments that result in its being culled although it may not be possible to determine a superlative animal without an examination of its offspring. An animal can be part of a regular herd until it is proved unfit. It can be held for the purpose of putting it to use even though not yet used.

Culling of immature or unused animals should be justified on grounds of nonconformity, failure to meet standards, sterility or the like. The intention to place animals in a breeding or dairy herd should be evidenced by records listing data on each animal which will support qualification.

"Use" Test

Revenue agents sometimes tried to determine the holding of livestock for breeding purposes by a test of actual "use." The government sometimes argued that the real test of the taxpayer's purpose is what he did with the animal; for example, if an animal had not been used for breeding purposes, but was shown and sold, the government claimed it was held for sale.

Every farmer and rancher has the burden of proof to show that his animals were held for draft, breeding, dairy, or sporting purposes. When the cattle owner is unable to show that his animals were held for breeding purposes, it is easy for the government to say that the best test of "held" is actual "use."

One should not confuse breeding with "used for breeding purposes." The fact that animals are bred before sale does not prove they were used for breeding purposes. It may be common practice to breed all females as soon as possible. In addition, the fact that heifers bred to excellent bulls and of proved fertility bring better prices is entirely consistent with the holding of animals for sale.

Age Test

Whether an animal is "held for draft, breeding, dairy, or sporting purposes" is purely factual and every livestock owner must establish a consistent practice and be prepared to prove it. If he does not regard an animal as part of his herd until it has been tested by its progeny, he has set his own pattern and the animal is not a part of his herd until it has produced or at least until it is old enough to have produced. This is the essence of the "age" test—a rule of thumb to determine whether the animal is old enough to have proved itself a part of the herd. It is not the government or the court that says that "use" or "age" is the test; the cattle raiser determines it. The decision is based on the facts of the particular farmer's or rancher's operation.

Thus, the taxpayer must make the age test unnecessary if he is to succeed. A pattern of early cull and selection, selective breeding practices, registration, disposal based solely on judgment of unfitness for the prospective use, and the absence of regular sales and advertising help to prove the taxpayer's case. Records are invaluable to show the nature of a carefully managed breeding or dairy herd.

The leading cases generally concerned taxpayers who were building exceptional herds, but the facts of such cases can guide every farmer. It is important that breeding and sale herds are physically separated, that breeders receive preferred food and treatment, and that animals in the breeding herd are not held out for sale and are seldom sold to other cattle breeders. Every farm operation is different, but every farmer can establish the facts of his own herd by establishing a fixed pattern of operation.

Primary Purpose

The statute requires that the animals be held for draft, dairy, breeding, or sporting purposes but it does not say that the purpose must be the primary purpose or even an equal purpose. Regulations under the prior law said that an animal held for ultimate sale to customers in the ordinary course of the taxpayer's trade or business may be considered as held for draft, breeding, or dairy purposes, but the negligible use of an animal for draft, dairy, or breeding purposes, or so using the animal as an ordinary or necessary incident to some other purpose, does not make it held for draft, dairy, or breeding purposes.

However, as a practical matter, if livestock is currently held for dairy, breeding, or sporting purposes and also for sale to customers in the ordinary course of the owner's trade or business, the issue is going to turn on which is the primary purpose. If the animals are currently held for breeding or racing purposes but might later be sold, the case is easier. For example, the mere fact that brood sows used for a single litter are held for ultimate sale in the ordinary course of business does not prevent them from being Section 1231 assets. A farmer who actually used his animals for draft, dairy, breeding, or sporting purposes is not precluded from applying the benefits of the law to the sale of such livestock by later selling them in the ordinary course of business.

Nor is property taken out of the class of "used in the business" simply because the business use is abandoned for a period before a sale is made. Consequently, conditioning a brood sow, or an abandoned dairy cow, in the feed lot for a period before sale does not prevent the application of Section 1231 to the proceeds of the sale of the animal. Nor does a farmer holding a closing-out sale become engaged in the business of selling dairy cows and brood sows if he hires an auctioneer, pays commissions, advertises the animals for sale, and makes other efforts to get the best price obtainable for the animals he had used in his business.

In cases in which all or a large part of a farmer's stock is sold, his position is greatly strengthened if he can show a special reason for the sale such as drought, sterility, sale of farm, retirement from business, shortage of hired help, or other special factors.

In the past, beef and dairy cattle created the most difficulty because they breed late. A heifer is over two years old before she is a proven producer. The time between the end of the twelve-month holding period and calving was the period which raised qualification questions. The extension of the holding period to two years should eliminate the vast majority of squabbles over cattle.

The holding period was not extended for sheep, and they will continue to raise these problems since a ewe is around two years old before lambing. Intent must be evidenced by good records, when it cannot be evidenced by use.

Sows held for breeding purposes do not cause much trouble because they are bred at around eight months of age and will have farrowed by the time the holding period of twelve months has expired.

Treasury Regulations

Treasury regulations use the following examples to illustrate when an animal is held for draft, dairy, or breeding purposes.

> **Example 1.** An animal intended by the taxpayer for use by him for breeding purposes is discovered to be sterile or unfit for the breeding purposes for which it was held, and is disposed of within a reasonable time thereafter. This animal is considered as held for breeding purposes.
>
> **Example 2.** The taxpayer retires from the breeding or dairy business and sells his entire herd, including young animals which would have been used by him for breeding or dairy purposes if he had remained in business. These young animals are considered as held for breeding or dairy purposes. The same would be true with respect to young animals which would have been used by the taxpayer for breeding or dairy purposes but which are sold

by him in reduction of his breeding or dairy herd because of, for example, drought.

Example 3. The taxpayer in the business of raising hogs for slaughter customarily breeds sows to obtain a single litter to be raised by him for sale and sells these brood sows after obtaining the litter. Even though these brood sows are held for ultimate sale to customers in the ordinary course of the taxpayer's trade or business, they are considered as held for breeding purposes.

Example 4. A taxpayer in the business of raising horses for sale to others for use by them as draft horses uses them for draft purposes on his own farm in order to train them. This use is an ordinary or necessary incident to the purpose of selling the animals, and, accordingly, these horses are not considered as held for draft purposes.

Two other illustrations in the regulations point out that the farmer who raises and sells breeding cattle holds the animals for sale, not for breeding (even though he breeds them to test them), and that the farmer who runs a feeder operation and purchases cows with calf does not obtain capital gain treatment on the ultimate sale of the calves.

There are also two examples involving race horses. They point out that true "culls" which have raced or been trained for racing may be Section 1231 assets.

Loss Benefit

The existence of the 24-month holding period for cattle and horses and twelve months for other livestock to qualify under Section 1231 provides an incidental benefit for farmers. Bulls, boars, brood sows, and similar animals are often bought by a farmer for temporary use and sold at a loss before the end of the minimum holding period. The loss is fully deductible and it does not have to be aggregated with Section 1231 gains.

Farm Income Tax Manual, §§ 315 to 325, 329.

§ 3.15 LIVESTOCK HELD FOR OTHER PURPOSES

Livestock not held for draft, dairy, breeding, or sporting purposes may be assets "used in the business" entitled to Section 1231 treatment. The holding period is more than six months, in 1976, nine months in 1977, and one year thereafter.

Poultry, fish, frogs and reptiles should be "livestock" for this purpose.

Livestock raised or used by the farmer may be property used in the business, although not held for draft, dairy, breeding, or sporting purposes. For example, a "wether" (an altered male sheep), if owned more than six months in 1976, nine months in 1977, and one year thereafter and held for the purpose of producing wool, meets all the tests of the statute. So does the business watchdog or sheepherder.

Angora rabbits held for the same periods and for the purpose of clipping their fur would seem to meet the general tests.

A fighting cock, a carrier pigeon, a jumping frog, a snake charmer's cobra and many other animals could be used in the taxpayer's particular business for other than draft, breeding, or dairy purposes. Their purpose may very well be "sporting," but they are excluded from the definition of livestock which has an elongated holding period by statute and regulation. Thus they should qualify for capital gain treatment if held for the proper period.

Farm Income Tax Manual, § 330.

§ 3.16 UNHARVESTED CROPS SOLD WITH THE LAND

Unharvested crops on land which is used in the business and held more than the minimum period, if crop and land are sold or exchanged (or involuntarily converted) at the same time and to the same person, are Section 1231 assets. Before 1977, the minimum period was six months; in 1977, it is nine months; in 1978 and later years it will be one year.

To qualify for capital gains treatment the growing or unharvested crops and land must be sold or exchanged together under the following conditions:

(1) The land must be held for more than the minimum period of six, nine, or twelve months (depending on the tax year involved). It is not necessary that the crop itself be held for any specified period or that it reach any particular state of maturity.

(2) The land must be used in the seller's farming business.

(3) Crop and land must be sold or exchanged at the same time.

(4) Crop and land must be sold or exchanged to the same person.

(5) No right or option to reacquire the land directly or indirectly may be retained by the seller (except a mortgage or other security contract).

Condemnation or seizure or sale under threat of condemnation is the same as a sale. Sale by a tenant of his leasehold with growing crops does not qualify.

Capital gains on the sale of unharvesteded crops are denied if there is a right to reacquire the land directly or indirectly. Most transactions involving reacquisition can be easily recognized, but some are less clear. For example, the government was sustained in its denial of capital gains on the sale of unharvested crops by a farmer to his controlled corporation on the ground that by reason of his complete control of the corporation, there was an indirect retention of the right to reacquire the land. But this control must be personal to the taxpayer and does not include the right to manage his wife's one-half of the community property.

No Deductions for Raising Crop

The seller must treat the costs of producing the crop as a capital investment rather than as an expense. In the year of sale of unharvested crops with the land, the farmer may not deduct the costs of raising the crop. This includes all items attributable to the crop, including depreciation; the government denies current depreciation on buildings and machinery if the entire farm and crops are sold. These expenses are added to the cost or basis of the land and reduce the gain or increase the loss.

It is possible that the farmer may have already claimed expenses as an offset against ordinary income in the previous tax year. For example, fall plowing of the field, seed costs, fertilizer costs and other possible expenses attributable to the current year's crop might have been deducted on the tax return for the previous year. In such cases, a recomputation on an amended return should be made, taking into account that expenses paid or incurred in that tax year for growing the crop must be eliminated.

Excess Deductions Account

Unharvested crops are farm recapture property which will produce ordinary income on sale to the extent of any balance in a farmer's "excess deductions account."

Farm Income Tax Manual, § 331.

§ 3.17 CUTTING OF TIMBER TREATED AS A SALE

A taxpayer who owned or had a contract right to cut timber prior to the beginning of the year of cutting may elect to treat the cutting of the timber as a Section 1231 sale or exchange.

The timber or contract right must have been held more than six months before 1977, more than nine months in 1977, and more than a year in 1978 and later years.

See § 3.18 for capital gains from timber royalties.

A timber owner may elect to treat the cutting of timber as a sale or exchange under the provisions of Section 1231. The election is available not only to an owner, but also to a person who had a contract right to cut the timber for more than the minimum period prior to the beginning of the tax year in which the timber is cut. If the cutting is in 1977, the timber or contract right must have been held more than nine months before 1977; in 1978 and later years, this period becomes more than one year.

The practical effect is to give the owner or cutter capital-gain treatment on the increase in the value of the timber to the year of severance. The election is made in the tax return for any year in which the taxpayer wants to start it, simply by computing the tax under these provisions.

If the election is made, the difference between the actual cost or the adjusted basis for depletion and the fair market value of the standing timber on the first day of the year in which it is cut is a Section 1231 gain or loss. Under normal circumstances an election would not be made if a loss is anticipated, but once the election is made it applies to all timber owned or subject to contract rights, whether acquired before or after the election.

The timber must be cut for sale or for use in the taxpayer's trade or business. If the election is claimed by one who is not the owner but who has only a contract right to cut timber, he must have an unrestricted right to sell the timber on his own account or use it in his own trade or business. It is sometimes difficult to determine from the nature of a cutting contract whether the taxpayer has the right to cut, remove and sell timber for his own account or whether he acts only for the account of the owner of the timber.

Adjusted Basis for Depletion

The computation of the "adjusted basis for depletion" requires allocation to the timber of a fair portion of the total price paid for the real estate. This amount is then reduced to depletion units (e.g., cords or board feet) so that when the timber is cut, the unit costs can be determined.

> **Example:** A farmer paid $50,000 for land on which was standing timber. Of this price, $25,000 was fairly allocated to the standing timber. With the help of an expert in such matters, he determines that he has 500,000 board feet of timber available. His cost basis is five cents per board foot or $50 per 1000.

A more rough-and-ready method might be used; for example, if the farm owner cut one-tenth of his standing timber during the tax year, his depletion basis for the timber is one-tenth of the $25,000 which he paid for all the timber on the land.

Subsequent Sale of Cut Timber

Having made the election to treat the cutting as a sale, the timber owner usually has another transaction to report when he makes a sale of the timber either in that form or in the form of some product manufactured from it. The profit or loss reportable is the difference between the price received and the new cost basis (the fair market value on the first day of the year in which the timber was cut). This is ordinary income or loss not subject to Section 1231 treatment.

If a taxpayer is governed by the election to treat cutting as a sale and uses the cut timber, for example, to build a barn, he would use the fair market value on the first day of the year as the cost of the lumber used in the barn. This is because the election specifically applies both to sales and "use in the taxpayer's trade or business."

Farm Income Tax Manual, § 328.

§ 3.18 ROYALTIES ON TIMBER, COAL, AND IRON ORE

Timber owned for more than six months (nine months in 1977; one year in 1978 and later years) which is disposed of under a royalty contract in which the owner retains an economic interest in the timber is treated as a sale under Section 1231. Similar benefits are extended to royalties from the disposal of coal and iron ore deposits.

If the owner of timber disposes of it under any form or type of contract in which he retains an economic interest in the timber and if he has held the timber before disposal, for more than six months in 1976, nine months in 1977, and one year thereafter, he is treated as if he made a sale of the timber under Section 1231. This is the usual form of "cutting contract," which pays the owner a royalty on timber cut. His gain or loss from the timber for each year is the difference between the amount received in the year and the adjusted basis for depletion of the timber sold.

The adjusted basis for depletion is the cost of the timber as allocated from the purchase price for the land and timber, less any adjustments necessary for timber already removed. The basis for the total amount of timber is divided by the number of units of timber to obtain a depletion unit. Depletion takes place when the timber is cut, so the depletion units are charged each year against the income from the units of timber cut and paid for under the contract.

The royalty provisions are mandatory. There is no choice or election as there is in the privilege of treating the cutting of timber as a sale. See § 3.17.

The date of disposal of the timber for determining whether it has been held long enough to qualify is the date of cutting. However, if the taxpayer is paid before the timber is cut, he may elect to treat the date of payment as the date of disposal. If the payment is made before the holding period ends, the election will remove the royalties from Section 1231 and they will be treated as ordinary income.

Advance payments and bonuses are also treated as Section 1231 income so long as the payments are for timber cut and the holding period is satisfied. If such advance payments are made, the timber is never cut and the contract expires, the bonus or advance royalty is ordinary income.

Right to Cut

These capital gain provisions apply only to a person who owns an interest in the timber, including a sublessee and a holder of a contract to cut timber. The taxpayer must have a right to cut the timber for sale on his own account or for use in his own trade or business.

Evergreens are timber for this purpose if they are more than six years old at the time of severance and are sold for ornamental purposes such as Christmas trees. Much interest has developed among farmers in the combination of the capital gain provision for Christmas trees and long-term forestry arrangements.

Expenses of Sale

Expenses directly related to the disposal of timber are treated as reductions of the sales price. Such costs include advertising for sale, marking for cutting, scaling, measuring, supervision of performance, foresters' fees, and other associated costs.

Christmas Trees

The Internal Revenue Service has long been unhappy about investors in Christmas trees who deduct the costs of raising the trees and claim capital gains on the ultimate sale under Section 1231. The Service demands that many of these costs be capitalized into the trees or the land.

While the Internal Revenue Service once insisted that expenditures incurred for planting, basal pruning, stump culture, and shearing had to be capitalized, it has retreated, and consistent with court decisions, now allows current deductions for the cost of pruning or shearing. Both the Service and the courts agree that expenditures incurred for silvicultural practices, such as weeding, or cleaning, or noncommercial thinning, are deductible expenses.

Preparation of the site including brush removal, cost of seedlings and labor and tool expense, including depreciation of equipment used in planting, are capital expenditures. Brush removal work performed a year or two after planting is considered proximately related to the establishment of the seedlings and its cost should be capitalized. The labor and tool expense includes all costs involved in planting the seedlings including all the amounts expended for transportation, supervision and labor, equipment rental and depreciation of owned equipment and tools used in connection with the planting. A portion of the depreciated cost of the equipment is added to the basis of the seedlings proportionate to the use of the equipment or tools in planting as compared with the use of such equipment in other activities.

The costs of land improvements such as road grading, ditching and firebreaks should be capitalized into the land; they should not be added to the basis of the trees. If the improvements have a determinable useful life they may be depreciated; otherwise the cost is added to the basis of the land. If there is only a right to cut, and no land ownership, the cost is amortized over the life of the contract.

Coal and Iron Ore

When coal or iron ore owned for more than the required period is disposed of by the owner under a royalty contract in which he retains an economic interest in the deposit, the owner treats the difference between the amount realized from such disposal and the adjusted depletion basis as a gain or loss on the sale. This gain or loss is specifically entitled to the preferential treatment provided in Section 1231.

Farm Income Tax Manual, §§ 326, 327.

§ 3.19 PROPERTY USED FOR BUSINESS AND PERSONAL PURPOSES

If property is used partly for business purposes and partly for personal purposes, it is considered partly a Section 1231 asset and partly a personal asset.

When a farmer keeps property, such as a car, solely for business use, it should be treated in all respects like any other machinery or equipment in his business. It is a Section 1231 asset because it meets all the general tests set up in the statute.

When an asset used partly for business purposes and partly for personal purposes is sold, the business and personal portions must be separated. It may result that the business portion which has been reduced by depreciation is sold at a taxable gain subject to recapture of depreciation and investment credit, and the personal portion is sold at a loss which is not deductible.

See § 5.8 on trade-in of property used only partly for business purposes.

Farm Income Tax Manual, § 332.

§ 3.20 CAPITAL GAINS ON NONBUSINESS ASSETS

Capital assets may be property held for the production of income, such as stocks and bonds, in which case gains and losses are recognized, subject to limitations on deductions of capital losses.

Other capital assets are personal assets in which case gains are recognized but losses are generally not deductible.

Stocks, bonds, and any other kind of property held by a farmer for the production of income or for investment, but not used in the farm business, are capital assets and the sales of such assets will produce capital gains or losses. These sales should be reported on Schedule D.

Household goods, and other personal items are capital assets in the hands of the farmer. When the farmer sells personal assets at a gain, it is taxable in the same manner as gains on other capital assets. If the sale results in a loss, the loss is not deductible because it is not incurred in a trade or business, or on property held for the production of income. See § 4.21.

It is not likely that the farmer selling household goods will realize a taxable gain. The original cost of furniture, stoves, radios, television sets and the like, remains the farmer's basis for computing gain or loss, and ordinarily used household goods do not sell for more than they originally cost.

However, losses on household goods used in the business, such as furniture for the hired man's room, refrigerators, coolers or freezers used to keep products held for sale, fall under Section 1231 because the property is used in the farm business.

Only if there is an involuntary conversion, i.e., casualty, theft or condemnation, will gains or losses on capital assets not used in the business be aggregated with Section 1231 transactions. See §§ 3.12, 3.24, 3.25.

See § 3.12 for the discussion of long- and short-term gains or losses and the method of reporting these transactions.

Sale of Life Interest

Prior to 1969, a life tenant could sell his life interest recognizing gain only to the extent of the difference between the amount realized and a prorata share of the basis of the property in which he held the interest. Now, however, unless the life tenant sells his interest along with the remainderman, the life interest will have no basis. Thus, the full amount received will be gain, usually capital gain, which has the effect of converting the periodic ordinary income rents or profits into a lump-sum capital gain.

Farm Income Tax Manual, § 336.

§ 3.21 COMMODITY AND LIVESTOCK FUTURES

Commodity and cattle futures purchased for speculative purposes are capital assets. If futures are purchased for true hedging purposes, the gains or losses resulting are ordinary gains or losses. The holding period for futures is still "more than six months."

Ordinarily when a farmer deals in commodity futures or carcass and live cattle futures, he has a capital gain or loss. These transactions will be reported on Schedule D, and if there is a loss, the loss is subject to the limitations on capital losses. See §§ 2.4-2, 3.12.

Futures contracts are treated as "rights in a specific commodity." As such they may be, for federal income tax purposes, capital assets except where they are entered into as "hedges." Unlike other capital assets, the holding period for long-term capital gain treatment on commodity futures contracts remains at "more than six months." It was not changed by the 1976 Act.

However, a farmer engaging in true hedging transactions as a protection against price fluctuations has ordinary income or loss from such transactions. This means that gains are fully taxable and losses fully deductible. For example, a livestock feeder might hedge against price rises of his feed by buying corn futures and holding them until it becomes certain that there will be a good crop so as to assure an adequate supply of feed at reasonable cost. By this time the price of the futures will also have come down so the sale of the futures will almost surely involve a loss to him. Had the purchase of the futures been a purely speculative venture, the loss would not be a full offset like other business expenses but would be treated as a capital loss. However, as a hedging transaction it would be a fully deductible business expense.

A skyrocket of grain and livestock prices could force "hedgers" to buy out of short-sale positions originally entered into as hedging transactions. To the extent these trades looked like "hedges" at the time they were entered into, the resulting losses should be fully deductible.

For example, a soybean grower might want to lock in what he thought was a good price by selling "short" futures contracts in soybeans and holding those contracts until he could sell his crop. By that time, the price of soybeans might have risen sharply so that, in order to buy out of his short position, he was certain to have incurred a loss. Had this transaction been a purely speculative venture, the loss would not be a full offset like other business expenses, but would be treated as a capital loss, subject to the limitations on capital losses. As a hedging transaction, however, it would be a fully deductible expense of his business.

It can be seen that, where speculative "short" sales are involved, seldom, if ever, does a long-term capital gain or loss result. The holding period starts when the futures contract used to cover the short sale is acquired, and that period is invariably one day at the most, thus falling far short of the required six-month period. The result from such speculative short sales is either a fully taxed short-term gain or a short-term capital loss, still subject to the capital loss limitations.

True Hedge

To qualify as a "true hedge" and thus avoid the capital loss limit:

(1) The taxpayer must trade futures in substantially the identical commodity which he raises, produces, or uses, and

(2) The purpose of the trade must be price insurance, not speculation.

In a true hedge, speculative risks are eliminated. The purpose of buying the futures contract must be price insurance and the transaction must be a business deal involving actual merchandise.

The taxpayer should buy regularly the identical commodity involved in his business. Corn and wheat through custom of the trade might be used as hedges against each other. Feed and cattle might also be used to hedge.

One farmer who raised corn and soybeans claimed his considerable trading in the commodity market was a hedge. But the Tax Court didn't buy his argument and called his losses "capital losses" since his trades involved crops which he did not raise, were frequently on the long side and his short position was regularly in excess of his possible production. The Court found no connection, interaction or relationship between the farm operations and his dealings in the futures markets.

In addition to the problems this farmer ran into, the fact that a short sale is closed out before the cash product is sold likewise detracts from its qualification as a hedge. That in itself, however, is not likely to spoil an otherwise legitimate hedge.

Most important is the "business insurance" aspect. That is, there must be a direct relation between the product which is the basis of the taxpayer's business and the future in which he deals for protection.

Further, there must be a direct relation between the prices of the product and the futures commodities traded. The prices of the futures must move up and down with the prices of the products of the business.

Usual Practice

Like many tax issues, the taxpayer's purposes for futures trading must be inferred after the fact from his course of dealing. It's also likely that IRS will look at the taxpayer's business practice over a period of years as some evidence of his intention in entering the market in the year in question.

But there is cause to be wary: in practice, commodities brokers advise that the dealings of many so-called farmer-hedgers resemble closely those of the commodities speculator.

Farm Income Tax Manual, § 337.

§ 3.22 SAND, GRAVEL, ROCKS, AND MINERALS

Ordinarily, sales of sand, gravel, rocks, and minerals produce ordinary income when sold under royalty or licensing agreements and result in capital gains when the deposit is sold "in place."

Deposits of sand, gravel, rock, or minerals on the farmer's land, if sold for a lump sum will be more likely to qualify as capital assets than as Section 1231 assets. Such materials are not likely to be used extensively in the business of farming (although the farmer may spread some of his gravel on roads or barnyards) and, therefore, they do not seem to meet the tests applicable to Section 1231 assets.

If the farmer is engaged in the business of buying or selling whole deposits of sand, rock, gravel, or minerals, he would have ordinary gains and losses on items held for sale and not capital gains and losses. If the farmer is engaged in the business of renting or licensing or receiving royalties on the removal of such materials as sand, gravel, and rock from his premises, income from royalties is ordinary income and fully taxed subject to a deduction for depletion. See § 5.14.

Whether the proceeds of the sale of such deposits are to be treated as ordinary income or capital gain generally depends on the nature of the transaction. Usually, it can be determined on the basis of a sale for a lump sum or a lease for royalty payments. However, if the farmer divests himself of all of his interest in the deposit (or a divided or undivided interest in the deposit), the transaction is a sale (and not a lease or license) resulting in capital gains even though the price is based upon production.

Where the farmer retains an economic or underlying interest in the deposit, he will generally be regarded as leasing the tract for the extraction of the sand, gravel or other deposit as a commodity. He will receive ordinary income and be entitled to a depletion allowance against the units sold. If the payment is dependent on extraction of the deposit and the farmer must look to the extraction for his money, he has retained an economic interest.

Although the theory of the distinction between ordinary income from royalties and capital gains from a sale of minerals "in place" is easy to state, it has been an incredibly difficult line for the courts to draw. In the most recent cases, the courts have been almost unanimous in finding that the sale of sand, gravel or other deposit for a fixed price per unit extracted produces ordinary income unless the buyer has an obligation to remove all or a specified portion of the deposit. One case did, however, find a capital gain because the dirt was to be taken from an area that was to be converted into a pond. The amount of dirt sold could be measured by the specifications for the pond; thus, the contract was for the sale of dirt "in place."

Farm Income Tax Manual, § 338.

§ 3.23 POSTPONEMENT OF GAINS AND LOSSES

Ordinarily, gains and losses are reported by a farmer at the time that they occur or are realized. In a few circumstances, gains and losses are not recognized for tax purposes in the sense that the reporting of the gain or loss is postponed until some time in the future. In some cases, the farmer has a choice; in others, the statutory postponement is mandatory.

The most important postponement provisions are discussed in the next sections:

§ 3.24—Involuntary Conversion and Reinvestment

§ 3.25—Condemnation and Reinvestment

§ 3.26—Installment Sales

§ 3.27—Exchanges of Property

§ 3.28—Sale and Repurchase of Residence

§ 3.24 INVOLUNTARY CONVERSION AND REINVESTMENT

Gains and losses from involuntary conversions of assets used in the business and held more than the long-term holding period are aggregated with other gains and losses from sales or exchanges of Section 1231 assets in determining whether the over-all result is a net gain or a net loss.

Gains and losses (if the losses are deductible) from involuntary conversions of capital assets held more than the long-term holding period are included in the computation of gains and losses from Section 1231 transactions.

The long-term holding period is six months before 1977, nine months in 1977 and one year thereafter.

A special rule excludes gains and losses from casualties and thefts if there is a net loss.

Involuntary conversions include the destruction or sale of livestock because of disease, the sale of livestock because of drought and the destruction of crops.

Gains on involuntary conversions are not recognized if the farmer purchases other property similar or related in service or use, within the time permitted by law.

Condemnation is an involuntary conversion but there are a number of special rules which are discussed in § 3.25.

An involuntary conversion occurs when property is converted into other property or money as a result of its destruction in whole or in part, theft, seizure or requisition, condemnation, or sale under threat or imminence of condemnation.

Insurance proceeds, condemnation awards, or other compensation will create a gain if the amount received is in excess of the adjusted basis of the property destroyed, stolen or condemned. If no compensation is received, or if the compensation is less than the basis of the property, there will be a loss.

Gains and losses resulting from the involuntary conversion of property used in the trade or busines or capital assets are treated as gains and losses resulting from Section 1231 transactions and are included in the aggregation of such gains and losses to determine whether there is a net gain or a net loss. To fall into the Section 1231 category, both the business assets and the capital assets must have been held over (1) six months until 1977, (2) nine months in 1977, and (3) one year in 1978 and later years (draft, breeding, dairy and sporting livestock for twelve or 24 months). It should be noted that the only time that an investment or personal capital asset is included in the Section 1231 aggregation is when the gain or loss arises from an involuntary conversion. See § 3.12.

A new wrinkle in the computation of Section 1231 gain or loss was added in 1969. The gains and losses from casualty or theft are separately aggregated. If there is a net gain, all of the transactions are treated as falling under Section 1231. If there is a net loss, none of the transactions fall under Section 1231 and the net loss is, in effect, fully deductible.

Nonrecognition of Gains

If property is substituted in kind as a result of an involuntary conversion, there is no gain recognized if the substituted property is similar or related in service or use. For example, if the state condemned property and conveyed a similar property in exchange, there is no election to report the gain. The transaction must be treated as a non-taxable exchange and the gain is not recognized.

If money or unlike property is received as a result of an involuntary conversion, gain will be recognized unless a special election is made. The taxpayer may elect not to recognize the gain if he purchases within the replacement period either similar property or a controlling interest in a corporation owning similar property, the cost of which equals or exceeds the proceeds from the old property. If the replacement property cost is less than the proceeds received, the gain will be recognized to the extent of the unexpended proceeds. Only the recognized gains on involuntary conversions are to be included with the Section 1231 transactions.

Losses

Losses on involuntary conversions are recognized. The postponement elections do not apply to losses. If the loss is properly deductible (see § 4.21), it will be fully deductible or it will be a Section 1231 loss. Losses on assets used in the business and capital assets resulting from involuntary conversions which are not casualties or thefts are included as Section 1231 items if the assets were held for more than the long-term holding period. This means that the losses offset Section 1231 gains which otherwise would have been treated as capital gains.

Losses on involuntary conversions which arise from destruction or theft are first included in the casualty and theft gain and loss aggregate as noted above. Thus, casualty and theft losses will either offset gains from casualty and theft gains or will be fully deductible.

Even though property destroyed or stolen is not converted into other property or money (so that there is a simple loss instead of a conversion), the losses must be treated as involuntary conversions and aggregated with other casualty and theft gains and losses. However, if the property was held for less than the long-term period, the loss is fully deductible. See § 3.12.

On the involuntary conversion by casualty or theft of a capital asset held longer than these periods, an allowable loss is fully deductible if there are no casualty or theft gains.

Period for Replacement

To take advantage of the nonrecognition-of-gain provisions, the farmer must make the replacement within two years after the close of the first taxable year in which any part of the gain upon the conversion is realized. He has three years where business or investment real estate is condemned. Upon application to the District Director, he may obtain an extension of either of these time periods.

Execution of a contract to purchase within the time period is not sufficient. The "purchase" occurs only when the benefits and burdens of ownership pass to the buyer, normally possession.

The Internal Revenue Service argues that when the proceeds of an involuntary conversion are received by the decedent before his death and the proceeds are reinvested by the estate in what would ordinarily be replacement property, the nonrecognition-of-gain provisions do not apply. The courts have not sustained the Service in this position, however. Partners cannot replace property converted in the hands of the partnership.

Replacement Property

The Internal Revenue Service has long taken the position that "similar or related in service or use" means that the property acquired must have a functional similarity to the property converted. Under this test, property was not considered similar or related in service or use to the converted property unless the physical characteristics and end uses of the converted and replacement properties were closely similar. For example, a harvester is not similar to a truck even though both are used in the business. Furthermore, the same relation must exist with the taxpayer. For example, a residence rented by a taxpayer cannot be replaced with a house to be occupied by him. See § 3.25 where both the condemned and the replacement property are business or investment, rather than personal, assets.

Livestock

Livestock destroyed by disease or on account of disease (including ordinary death losses from disease) or livestock sold or exchanged because of disease are treated as if they were involuntarily converted. Similar treatment is given to a farmer who sells draft, breeding, or dairy livestock because of drought.

Insurance proceeds, sale proceeds, or amounts received from the government for condemnation of diseased livestock may be reinvested in similar animals, and the taxpayer may elect to have the recognition of the gain postponed.

The replacement property must be functionally equivalent, i.e., dairy cows for dairy cows. The replacement privilege applies to all diseased livestock whether held for use in the trade or business or held for sale, if the animals would not otherwise have been sold at that time, and it is not limited to draft, breeding, and dairy livestock. The taxpayer must attach to his return a recital of the evidence showing that the livestock were killed or sold because of disease.

If the animals are not replaced, gain is recognized. Loss is always recognized. If the animals were held for sale, the farmer will report ordinary gain or loss on Schedule F. If the animals were used in the trade or business, the gains or losses will be reported as Section 1231 assets or not, depending upon the period of holding. See § 3.12.

The Internal Revenue Service holds that disease is not a casualty, so these gains and losses are not added to the casualty and theft aggregate.

Livestock Sold Because of Drought

Special treatment is given to a farmer who sells draft, breeding, or dairy livestock (regardless of how long they are held) because of drought. If he sells more of his livestock held for draft, breeding or dairy purposes than he would normally sell—solely because of drought—the excess number is treated as an involuntary conversion under Section 1231. No gain is recognized if the proceeds are reinvested in similar animals within the next two taxable years. Special permission can be obtained to extend the replacement deadline. This does not include sporting livestock.

If he does not replace the animals or if there is a loss, regular gain-and-loss rules apply.

The mere existence of drought conditions will satisfy the statute. There is no requirement that the taxpayer be unable to purchase feed for his cattle. Even though cattle were placed in a feed lot for two months prior to the sale for fattening, this did not change the breeding purpose for which the cattle had previously been held nor did it change the fact that the sale was made solely on account of drought.

There is a difference between these provisions— the "disease" provision covers all animals; the "drought" provision is limited to draft, breeding, and dairy animals. There is no qualifying holding period in either case.

There is also another exception for reporting gain on livestock because of drought. Gain on such sales in excess of the farmer's normal sales can be deferred until the following year. See § 3.2.

Destruction and Repurchase of Crop

The destruction by hail of a standing crop is an involuntary conversion. Gain from the insurance proceeds is not recognized where the proceeds are used to purchase another standing crop or a harvested crop within the replacement period. However, the use of the insurance proceeds to cover the cost of planting a new crop does not constitute the acquisition of replacement property.

No gain is recognized if the proceeds of condemnation of an orchard or grove are reinvested in uncultivated land and the planting of young trees. Costs of planting are replacement costs but costs attributable to young trees after the planting stage are not.

Gain or loss from crop insurance would not fall under Section 1231 because the crop was held for sale. But the trees would be Section 1231 assets because they are depreciable assets used in the business

See § 3.10 on postponement of reporting gain from the proceeds of crop insurance or government crop damage payments until the following year.

Reclamation Laws

Sales of property to conform to acreage limitations of reclamation laws are treated as involuntary conversions. Replacement property may be of "like kind" under the condemnation rules. See § 3.25.

Personal Assets

When the involuntary conversion involves a nonbusiness or noninvestment capital asset such as a personal residence, household goods or an automobile held exclusively for personal use and held for more than the long-term period, the gain on the conversion must be included in the computation of gains and losses from Section 1231 assets first passing through the casualty aggregate, if applicable. This is the only instance where personal assets are included with Section 1231 transactions.

If there is a loss on the involuntary conversion of a capital asset, it must be determined whether the loss is allowable. A loss which is not deductible at all is not resurrected by this rule. For example, a casualty loss to a residence is an allowable loss which is included in the casualty gain and loss aggregate under Section 1231. But a loss resulting from condemnation of the residence is a personal nondeductible loss, so it would not be included as part of the Section 1231 transactions, nor be deductible anywhere else.

Assuming that a loss deduction is allowable, a loss on an involuntary conversion of a capital asset held for less than the long-term period is not included with Section 1231 transactions. It will be reported as a short-term loss or a casualty loss. An allowable loss from an involuntary conversion of a capital asset held for more than the period will be a Section 1231 item or a fully deductible loss if there are no casualty gains.

The gain on the involuntary conversion of a capital asset will not be recognized if the taxpayer replaces the property under the regular nonrecognition-of-gain provisions.

Farm Income Tax Manual, §§ 339, 340.

§ 3.25 CONDEMNATION AND REINVESTMENT

Condemnation or sale under threat of condemnation of an asset used in the business or a capital asset held for more than the long-term period is an involuntary conversion which follows the general rules outlined in § 3.24.

Certain rules peculiar to condemnation are discussed in this section—replacement period, replacement property, severance damages, sale under threat of condemnation and easements.

"Condemnation" refers to the process by which property is taken for public use without consent of the owner, with just compensation. All condemnations of business property or property held for the production of income are involuntary conversions whether the transaction results in a gain or loss.

Loss sustained from the condemnation of property held for personal purposes is not deductible, but a gain will be taxable unless a replacement election is made. If the personal residence is condemned, the taxpayer may elect to postpone the taxation of any gain either as an involuntary conversion or under the sale and replacement of residence provisions. See § 3.28.

A sale under threat or imminence of condemnation is treated as a condemnation. Threat or imminence of condemnation exists when a property owner is informed by a representative of the condemning authority that it has decided to acquire the property and the property owner has reasonable grounds to believe that the property will be condemned if he does not sell. The sale does not have to be made to the condemning authority. It may be made to a third party and the proceeds reinvested without recognition of gain.

Measure of the Award

If the condemning authority retains out of the award an amount to satisfy special assessments levied against the remaining part of the property for benefits which accrue in connection with the condemnation, the amount of the special assessments is deducted from the amount of the gross award in determining the net condemnation award for purposes of gain or loss.

If the land condemned has a growing crop, no part of the award is allocated to the crop; the entire amount is considered to be received for the land. If the farmer is permitted to harvest the crop, any amounts received from the sale or other disposition of the crop will be ordinary income.

Any amounts used to pay off a mortgage or other lien on the condemned property are not deductible from the award, nor are they treated as an investment in replacement property. Such payments to a mortgagee or other lien holder are regarded as received by the farmer even though the money never passed through his hands.

Severance Damages

If severance damages are separately assessed and stated, they may be separately treated. The damages are applied first to the prorata share of expenses, then to costs of restoring the remaining property (rebuilding fences, roads, etc.), then to reduce the basis of the remaining land. If there is a special assessment, it would be charged to severance damages after expenses.

Severance damages must be stipulated in the contract with the condemning authority or specified in an itemized statement or closing sheet. Consequential damages and damages of a similar nature caused by public improvements, but which involve no actual taking of property, are treated in the same manner as severance damages.

Gain results if the amount of severance damages is in excess of the basis of the retained property. The Internal Revenue Service permitted a farmer who received severance damages on a partial taking to invest the award and the severance damages in the purchase of adjacent land.

Replacement Period

In the case of condemnation, the period during which the taxpayer may acquire replacement property begins at the earliest date of the threat or imminence of requisition or condemnation. It is possible, therefore, for the farmer to buy his replacement before the actual condemnation takes place. In such case the property must be held by the farmer on the date of the condemnation and the replacement must be property which he acquired by purchase or exchange. For example, it would not be sufficient if the taxpayer acquired similar property by gift or as an heir. The period of replacement ends two years after the end of the first tax year in which any part of the gain is realized, except replacement of business or investment real estate may be deferred until the end of the third year. An extension of time may be obtained from the District Director.

In condemnations under the Federal Declaration of Taking Act, title to the condemned property passes to the Government even though the final price has not been fixed, and the payment is made into the District Court. This fixes the date of sale and the year in which part of the gain is realized. Subsequent payments resulting from litigation or settlement will be income when received. If the condemnation procedure is not under federal rules, then the date of sale and the year in which gain is first realized will be determined under state or local law.

Replacement Property

If real property is used in a trade or business or held for the production of income or investment, it may be replaced with property of a "like kind" as well as property "similar or related in service or use." If the proceeds are invested in stock of a corporation, it must own property "similar or related"; "like kind" will not do. See § 3.24 for the meaning of "similar or related."

"Like kind" means property of the same general class or nature or character but need not be of the same grade or quality. For example, farms, urban realty, and improved and unimproved land are real property of a like kind, but a fee interest in a farm and a ten-year lease are not of like kind.

The use of proceeds from condemnation of land to build improvements, e.g. buildings, roads, water systems, on land already owned is not reinvestment in "like kind" property. Nor is the use of the award to pay off the mortgage on adjacent farm land.

If a farm is condemned and the proceeds are used to purchase an interest as tenant in common in a larger farm, the gain on the original farm may be postponed since "like kind" property is acquired. But if the proceeds from the involuntary conversion of the farm were invested in a farming partnership owning similar farm land, the gain would be recognized because a partnership interest is regarded as personal property—an unlike kind. Only real property held for productive use in the trade or business or for investment may be reinvested in property of like kind.

Condemned property was replaced without recognition of gain when acreage submerged under water was filled in. The farmer obtained the title from the governmental authority. The cost of the replacement property included expenditures for preliminary surveys, permits and approvals, fill materials (including transportation and handling costs), mineral rights, surveys of the filled in land, issuance and recording of the deed or patent, and any other payment directly related to the acquisition of the property.

In all cases of involuntary conversion, other than condemnation of real estate held for business use or investment, the replacement property must be similar or related in service or use; it must be functionally the same as the property converted. A business vehicle must be replaced with a business vehicle which must perform the same functions. A personal residence must be replaced with a personal residence, a dairy cow with a dairy cow. Further, it has to serve the same purpose for the taxpayer; for example, a condemned rental residence may not be replaced with a house to be occupied by the taxpayer.

The words "similar or related in use" were strictly interpreted in the past. For example, unimproved real estate was not similar to improved real estate. More recently, however, property used for a warehouse was similar to property used for a gas station since the taxpayer's relation to and functional use of the two properties was the same; he leased both properties to others. As a practical matter, much of the replacement problem is solved by the "like kind" rule on condemnation of business and investment real estate, while the more narrow rule applies to replacement of personal property and real estate held for personal purposes. On any involuntary conversion, other than condemnation, the "similar or related" definition applies to all property including real estate. See § 3.24.

In all cases, the replacement property must be acquired by purchase. Property acquired by gift or death will not qualify as replacement property to postpone the recognition of gain.

If the condemnation of part of a farm destroys the economic unity of the operating unit, the taxpayer may be forced to sell the portion which was not condemned and reinvest the sales price and the condemnation award in a new farm. Normally, the gain on the sold portion would have to be recognized, but the total proceeds may be reinvested without recognition of gain if the sale resulted from the destruction of the economic unit.

Easements

Condemnations of easements are involuntary conversions. When, for example, a state takes a strip of land for highway purposes, the land is involuntarily converted into money by reason of the condemnation proceedings or sale under threat or imminence thereof. The difference between the cost to the farmer of land actually taken and the award given in the proceedings is the gain or loss involved. The farmer treats the gain or loss involved as a Section 1231 gain or loss. See § 3.12.

If the farmer can continue to use the land involved in the condemnation proceedings, for example, when a utility secures a right of way across a farm for a power line, the award is then considered as a return of capital which reduces the cost basis of the land subject to the easement and does not have to be reported for income tax purposes at the time received unless the award exceeds the basis.

Damages for crop losses in connection with the grant of an easement are ordinary income. They are not part of the amount paid for the right of way which reduces the basis of the farm.

It is possible that the proceeds might be reinvested. For example, proceeds from the condemnation of an easement to flood land might be reinvested in dikes, levees and other measures to safeguard the property from the flooding that would follow from the installation of a dam by the condemnor. No gain would be recognized on the theory that the compensation for the taking of a right to use the property was reinvested in restoration of the usability.

Farm Income Tax Manual, § 339.

§ 3.26 INSTALLMENT SALES

A sale of property in which the payment of all or part of the purchase price is postponed will be reported in one of two ways. The usual rule requires that the gain or loss be recognized in the year of sale. A special election permits reporting the gain on some sales as the installments are received.

(a) Ordinarily, the cash and the fair market value of the obligations of the purchaser are regarded as the amount realized in the year of sale and the difference between the amount so realized and the adjusted basis of the property is the gain or loss to be reported in the year of sale.

(b) The special election to defer tax on gain until the actual payments are received applies to casual sales of personal property for more than $1000 and to sales of real estate if the payments in the year of sale do not exceed 30% of the selling price.

(c) Installment payments must bear at least 6% interest or a 7% interest element will be carved out of the purchase price.

(d) Savings from income averaging should be compared with the benefits from an installment sale.

(e) Special rules limit gain or loss on repossession of property by a seller.

Property is often sold under an arrangement by which the payments of the purchase price are made over a period of years. Under the ordinary rule, the transaction is treated as complete in the year of sale and the gain or loss is recognized and reported in the year of sale. The only exception to immediate recognition and reporting involves an election by the farmer to treat the sale under the "installment method" which permits him to report as gain a part of each installment as the payments are received.

(a) Ordinary deferred-payment sale

Under normal circumstances, the amount realized on a sale is the cash plus the fair market value of the obligations of the purchaser. The sale is complete and the income is realized when the title or the burdens and benefits of ownership pass to the buyer, whichever first occurs. If the obligations of the buyer are treated as equal to their face value, the total gain is recognized in the year of sale even though not all of the payments have been received. Since the gain has been recognized, subsequent collection of the obligations has no income tax effect.

However, if the fair market value of the obligations of the purchaser is less than face value, the gain in the year of sale is based upon the cash plus the fair market value of the purchaser's obligations. This means that if the obligations are ultimately paid off at the face amount, there is additional income to report at sometime in the future.

When the obligations of the purchaser are paid off, the seller has a gain measured by the difference between the fair market value of the obligation which he reported as income in the year of the sale and the total amount that he receives on the discharge of the purchaser's obligation.

Because the discharge of an obligation is not a sale or exchange, the excess that the seller receives on the discharge of the obligation is ordinary income. This is true notwithstanding the fact that if the purchaser had paid cash or if the obligations had been reported at their face value in the year of sale, all of the gain would have been capital gain.

There is a limited statutory exception to this rule. Unless issued at a discount exceeding 1/4% a year, the taxpayer would have capital gain and not ordinary income on retirement of a corporate or governmental obligation if it was a capital asset in the hands of the taxpayer.

Obligations Without Market Value

If the obligations of the purchaser have no fair market value, the ultimate gain or loss cannot be determined until the obligations are paid off. The seller applies all payments to recovery of the basis of the property and only when he is made whole does he recognize all subsequent payments as income when received. In this case, the sale transaction is never closed until all of the payments are made. Therefore, if the sale was a capital gain sale, the payments as they are received are capital gain, not ordinary income. If the sale results in a loss, the loss is deductible in the year in which the loss can be fully determined.

The Internal Revenue Service, except in very rare circumstances, finds it difficult to believe that the obligations of any purchaser have no fair market value. Some years ago, it was believed that the promise of a purchaser under a land contract to pay the purchase price had no fair market value. Presently, the government insists that the promise has value even though it may not be the full face value. At least one Circuit Court has agreed with the Service even though the value of the obligation was only one-half its face amount.

(b) Installment sale election

An election to report a sale on the installment method defers the tax on the gain until the money is received. A prorata part of each payment is treated as gain and the tax on that gain is reported in the year that the installment is received. Often the installment sale results in a lower over-all tax because the gain is spread over a period of years.

Installment sales made by farmers will usually involve either capital assets or business assets. Casual sales of any personal property (other than inventory-type property) for over $1000 or sales of any real property, are subject to the installment method of reporting.

No payment is necessary in the year of the sale, but payments in the year of the sale may not exceed 30% of the selling price. An installment-sale contract may provide for nothing in the year of the sale, but the balance must be payable in at least two installments in the future. A sale requiring a single future payment is not an installment sale.

The amount returned as income from an installment sale is that proportion of the payments actually received in the year which the gross profit on the entire transaction bears to the total contract price.

Installment-sale provisions apply only to sales that produce an income tax gain. If the sale results in a loss, the only spreading techniques available are the net operating loss carry-back and carry-forward (§ 4.18) and the capital loss carry-over (§ 3.12).

Interest received on the installment-sale contract or the note and mortgage or other obligation of the purchaser is ordinary income. The spreading provisions apply to the gains in the principal payments. If no interest is stated, it will nevertheless be carved out.

If the assets sold include property subject to depreciation recapture, the gain to be reported must first be allocated to the recapture of ordinary income. Thereafter, the balance may be allocated to other assets.

Lump-Sum Sale of Farm Assets

The installment sale of an entire farm for a lump sum under a single contract is not the sale of a single asset. It generally includes the sale of real and personal property which may be reported on the installment method and inventory items which may not be reported on the installment method. The sale price must be broken down to determine the amount received for each kind of property. The 30% test is applied to the part of the sale price received for property which may be reported under the installment method, and the $1000 test for casual sales of personal property is applied to the part of the sale price received for such property.

Payments in the Year of Sale

The law does not mean a "down payment" of 30%; it speaks of "payments in the year of sale." A farmer lost the benefit of the installment sale because he took a 30% down payment on March 1 and received an additional payment on principal before the end of the year. The seller may not receive or be entitled to more than 30% of the selling price in the year of sale. The value of any other property received or applied to the seller's benefit must also fall within the 30%.

The year of sale is the year in which the benefits and burdens of ownership pass from the seller to the buyer. Usually this is the transfer of possession. Title passage is irrelevant unless the title passes before any other transfer of benefit in which case the year of passage of title is the year of sale.

The statute demands that "payments" in the "taxable year of sale" not exceed 30% of the selling price. No payments in the year of sale are required, but if made, the aggregate may not exceed the 30% figure. There must be at least two payments over two taxable years, not including any earnest money or other payment prior to the year of sale, to have an installment sale.

Payments in the year of sale include a "down payment" or other payment made in a prior year. For example, suppose a contract was signed in November of 1976 providing for an immediate down payment or earnest money payment and transfer of possession of the farm on March 1, 1977 with a payment of 30% of the selling price to be made on March 1, 1977. "Payments in the year of sale" (1977) include the amount of the down payment or earnest money paid in the year prior to sale (1976).

The only safe method of handling the transaction is to be assured that all of the payments prior to and including the year of sale do not, in the aggregate, exceed 30% of the selling price.

Liabilities of the seller such as liens, accrued interest and taxes assumed and paid by the purchaser of the property in the year of sale generally will be treated as payments received in the year of sale. If these assumed liabilities, including a mortgage, exceed the seller's basis, the amount of such excess is treated as a payment in the year of sale even if the buyer does not pay them until a subsequent year.

Commissions and other selling expenses may not be deducted from payments received either for the purpose of keeping below the 30% figure or for determining the total "contract price" or the "selling price."

Cancellation of a seller's indebtedness to a purchaser as partial consideration for the purchase price is equivalent to the receipt by the seller of that part of the sale price. Therefore, such cancellation is to be treated as a payment in the year of sale in the application of the 30% rule. But the buyer and seller may agree in their contract that the indebtedness shall apply against later installments due and not against the payments in the year of sale.

If a sale contract provides for no interest or a low interest rate, the purchase price will be reduced by the amount of imputed interest under Section 483 of the Code. This reduction amounts to at least a 7% discounted interest rate. If the selling price is so reduced, the farmer may then find that the "payments in the year of sale" are more than 30% of the true selling price and thus lose his right to elect the installment method.

How to Report Sales

The income to be reported in any taxable year is that proportion of the payment which the gross profit bears to the "total contract price." The qualification for the use of the method is that not more than 30% of the "selling price" is received in the year of sale.

Selling price is the entire cost of the property to the buyer, including any mortgage on the property. If there is no mortgage on the property or if the sale is secured by a mortgage or contract from the buyer to the seller, the "total contract price" is the same as the selling price.

If there is an existing mortgage taken over by the buyer, the contract price is the selling price reduced by the mortgage. In effect, the "total contract price" is the seller's equity in the property.

> **Example:** A buyer purchases a farm for $50,000. He pays $10,000 in the year of sale, assumes a mortgage of $25,000 and agrees to pay $5000 per year for three years. Interest is 6% or more. The farm has a basis of $35,000 in the hands of the seller. The seller's gross profit would be $15,000 and the total contract price $25,000. The result would be that the "gross profit" ($15,000) bears the relation to the "contract price" ($25,000) of sixty per

cent. Thus 60% of the $10,000 payment and 60% of each $5000 installment is gain.

Disposition of Obligations

Disposition of the installment obligations results in immediate recognition of all of the postponed gain. A sale, gift or other disposition, or the satisfaction of an installment obligation causes the postponed gain to be recognized. Even the assignment of the notes as security may be a disposition in some situations, particularly where the transaction lacks the usual characteristics of a loan.

This does not apply to transfer by death of the rights to payments under the installment obligations. If a farmer dies owning installment obligations, the executor, administrator or legatee can carry on and report income in the same manner as the decedent would have done if he had lived. The successor is entitled to an income tax deduction for state and federal death taxes paid on the deferred income.

Election to Use Installment Method

The option to use the installment method applies to each separate sale. The election is made by treating the sale under the installment method on the tax return in the year of sale.

Even though the taxpayer receives no payment in the year of sale, nonetheless an installment sale election must be made on the return for the year of sale. However, an inadvertent failure to report a sale has been held not to be an election not to use the installment method.

An installment election on the sale of an old residence may be filed after a decision not to buy a new residence. If the requirements of an installment sale are met and the original return was timely, the installment sale method may be elected on the amended return reporting the gain on the sale of the residence.

(c) Sales must bear interest

A farmer cannot increase the purchase price of his farm and charge no interest or a very low interest and thus convert ordinary interest income to capital gain sales price.

On deferred-payment or installment sales, the interest element will be carved out of the sale price and treated as interest. Unstated interest even arises on a sale at a loss; the loss is increased by excluding the unstated interest from the proceeds.

If the contract is in excess of $3000, and one or more payments are due more than a year after the date of sale, and the interest rate is below 6%, a portion of each deferred payment is interest. The law applies to all payments due more than six months after the sale.

Generally, the rule does not apply to private annuity transactions. It does not apply to a seller if his gain is not capital gain, but the buyer treats his "interest" as deductible. Nor does it apply to the buyer if carrying charges on the sale price are deductible.

Thus, on an installment or deferred-payment sale of real or personal property over $3000, there must be simple interest of 6% or the unstated interest will be determined and carved out of the purchase price. For the seller, the sales price is reduced by the unstated interest and he reports the interest as ordinary income. For the buyer, the purchase price (and, therefore, the basis of the asset purchased) is reduced and the unstated interest is deductible.

(d) Compared to Income Averaging

If income increases in any one year, it may be averaged over the current and prior four taxable years. The averagable amount is the amount by which the taxable income for the year (reduced by (1) lump sum distributions from retirement plans and (2) trust "throw-back" distributions) exceeds 120% of the average taxable income for the prior four years. The averagable income must be more than $3000. Gains on property sales may be averaged, and in some cases the tax on the entire gain will, when averaged, be less or about the same as the tax on the gain if paid in installments. However, if the taxpayer qualifies, the installments may also be averaged as they are received. Comparison requires some speculation about income in the future and the tax rates when installments would be received, but averaging is likely to be more nearly the same or better when the gain is small or the number of installments few. But remember even if averaging produces a smaller tax, all of it must be paid this year. Under the installment method, the tax is paid only as the money is received. However, if all the money is received in the year of sale, the installment provision cannot apply. Then only averaging is available, but the taxpayer also has the funds to pay the full tax. If the buyer can pay, say by getting a new mortgage, then the tax savings from averaging and from an installment sale are more comparable. In either case, some speculation is required about other income in the year of sale and in future years and also future tax rates. Generally, more will be known about the year of sale simply because it is the closest in time. But the only way to make this comparison is to compute the tax under each method and establish which appears to be the least.

(e) Repossessions of property

If the sale of real property gave rise to an indebtedness to the seller which was secured by the real property sold and the seller repossesses the property in satisfaction of the indebtedness, there is no loss recognized on the repossession and there is a limit on the gain that may be recognized. Gain is recognized to the extent of the cash and the fair market value of the property received prior to repossession (other than the obligations of the purchaser) less the gain on the sale already reported. However, the gain cannot exceed the amount of gain on the original sale reduced by the gain reported as income and the repossession costs.

On subsequent sale of the repossessed property, gain or loss will be recognized. The basis of the property is the basis to the seller of the purchaser's obligations which were secured by the property plus any gain recognized on repossession plus the repossession costs.

If the property repossessed is a residence which, on the original sale, qualified for nonrecognition of gain (either because a new residence was purchased or because the taxpayer was over 65, see § 3.28) and the repossessed residence is sold within a year of repossession, the resale is treated as part of the original sale transaction and the repossession rules do not apply.

On property other than real estate, the gain or loss is measured by the difference between the basis of the obligations and the fair market value of the property repossessed. Basis of those obligations is their face amount reduced by the amount of income which would be realized if the obligations were paid at face. The basis of the repossessed property is its fair market value at the time of repossession.

Farm Income Tax Manual, §§ 344, 345.

§ 3.27 EXCHANGES OF PROPERTY

Generally an exchange of property for other property is treated as a sale. Gain or loss is recognized in the amount of the difference between the adjusted basis of the property transferred and the fair market value of the property received.

However, if business or investment property is exchanged for property of a "like kind," no gain or loss is recognized.

Ordinarily all gains and losses are recognized whether property is sold, involuntarily converted, exchanged, or otherwise disposed of. The gain or loss is determined in the usual manner by computing the difference between the amount realized on the disposition and the adjusted basis of the property. Exchanges of property for other property, for merchandise or for services are treated as if the property were sold for the fair market value of the thing or service received. The gain or loss will be ordinary or capital depending upon whether the asset transferred was an ordinary asset or a capital asset.

Tax-Free Exchanges

There is no way that a farmer may sell a farm or other asset on his own volition and reinvest the proceeds in a similar farm or asset without recognizing the gain on the sale.

There are three special circumstances in which gain may be postponed. The first involves condemnation or involuntary conversion discussed at §§ 3.24, 3.25. The second is the sale and repurchase of a residence (including a farm residence) discussed at § 3.28. The third is the exchange (not a sale and repurchase) of farms or other "like kind" property discussed in this section.

The only voluntary action which a farmer may take to avoid the recognition of gain or loss is to exchange business property or investment property for property of a like kind. In effect, the new property acquired on the exchange is treated as a continuation of the original property in substantially the same form. For example, the trade of a tractor for a tractor, a dairy cow for a dairy cow, or a farm for a farm is a continuation of the original investment upon which neither gain nor loss would be recognized. The fact that the value of the property received is more or less than the basis of the property exchanged does not matter if the exchange qualifies as "tax free."

"Tax free" is something of a misnomer. The recognition of the gain or loss is postponed until some time in the future when the property acquired on the exchange is disposed of in a taxable transaction.

Property of Like Kind

To qualify for a "tax-free" exchange, property must be held for productive use in the trade or business or held for the production of income and must be exchanged for property of a like kind which is held for either purpose. Property is of a like kind if it is of the same nature or character although not necessarily of the same grade or quality. For example, a farm may be traded for city real estate, or a small truck for a big truck but not a truck for a boat, or a farm for stock in General Motors, or a tractor for ten acres of land.

The tax-free exchange provisions do not apply to exchanges of property held primarily for sale, nor to stocks, bonds, notes, or other forms of securities.

If the trade is solely one property for another of a like kind, no gain or loss is recognized and the new property takes over the basis of the old. If the farmer pays a cash boot, there is no gain or loss recognized and the basis of the acquired property is the basis of the old plus the cash paid.

Receipt of Boot

On the exchange, if the farmer receives property of a like kind and also cash or other property, he must recognize the gain on the transaction to the extent of the cash and the fair market value of the other property.

> **Example 1.** If a farmer traded forty acres which cost him $12,000 for thirty acres of other farm land worth $16,000 and received an additional $2000 in cash, he would have a gain of $6000—but only $2000 is recognized.

> **Example 2.** If he received the thirty acres and a truck worth $1500, the gain would be limited to the fair market value of the truck or $1500.

Basis of Property Received

The basis of the property received on a partially tax-free exchange is the basis of the old property less the cash received plus the gain recognized. In the first example above, the basis of the thirty acres would be $12,000 (old basis) less $2000 (cash) plus $2000 (gain recognized) or $12,000. In the second example, where no money was received, the basis of the thirty acres and the truck would be $12,000 (old basis) plus the gain recognized ($1500) or $13,500. That basis covers two properties, so it must be allocated. The non-like-kind property, the truck, is assigned a basis equal to its fair market value ($1500) leaving $12,000 for the basis of the thirty acres.

No Loss Recognized

If there is a loss on an exchange, it is not recognized. For example, if a farmer traded a tractor with a basis of $2000 for another tractor worth $1000 and received $500 cash, his loss of $500 is not recognized. The basis of the acquired tractor is $1500 (old basis of $2000 less the $500 cash received).

Disposition of Property

Upon disposition of the property acquired on an exchange, the postponed gain or loss will usually be reported. In the first example above there was a gain of $6000 on the trade of the forty acres of which $2000 was recognized. If the thirty acres is sold for its fair market value of $16,000, its basis is $12,000 and the remaining $4000 of gain will be reported. In the loss transaction above, if the acquired tractor is sold for its value of $1000, the $500 loss will be recognized because the tractor has a basis of $1500. If the disposition is by exchange for property of a like kind, this exchange will also be "tax free."

Where a taxpayer dies after 1976 and property passes to a new owner at death, it takes a basis equal to the decedent's. If gain or depreciation is involved, this basis may be adjusted for increases in value before 1977 and also adjusted for federal and state death taxes paid on increases in value after 1976. As so adjusted, the basis may not exceed fair market value at death, however, to the extent such adjustments increase basis, the postponed gain or loss will never be recognized. Where the decedent died before 1977, property took a basis equal to fair market value. It was not uncommon then for a farmer to trade his farm land for an apartment building to avoid the gain during his life, and then leave the property by will to a beneficiary who took a new basis for depreciation and sale equal to the fair market value at death.

Mortgaged Property

When the properties exchanged are subject to mortgages, or other liens or obligations which are assumed by the respective parties, the computations become somewhat more complicated but the basic principles apply. Generally the assumption of a mortgage by the other party is treated as the receipt of money on the exchange. If each party assumes the other's mortgage, "money" is received to the extent of the excess of the mortgage given up over the mortgage assumed.

Phony Trade-In

The Internal Revenue Service has challenged the separation of the sale of used equipment from the purchase of new to avoid an "exchange" of the old equipment for the new.

The Service says that it will look to what is actually done rather than to the formalities of the transaction or the declared purpose of the participants. In the usual case of the separate sale to and purchase from the local dealer, the transactions are reciprocal and mutually dependent. The farmer's acquisition of the new equipment from the dealer is contingent on the dealer's accepting the used equipment and granting a trade-in allowance equal to or in excess of its fair market value.

Other Problems

A number of special problems arise in connection with exchanges, some of which are discussed elsewhere. See § 5.2 on extra first year depreciation, § 5.8 on trade of property used partly for business purposes and basis generally, §§ 5.9, 5.10 and 5.11 on recapture of depreciation and investment credit and §§ 5.12 and 5.13 on recapture of land deductions and the "excess deductions account."

Farm Income Tax Manual, § 346.

§ 3.28 SALE AND REPURCHASE OF RESIDENCE

On the sale of a residence, gain is computed and recognized according to the usual rules with two tax-saving exceptions.

(a) To the extent that a taxpayer reinvests the proceeds of the sale or exchange of his personal residence in another residence within 18 months (or builds another residence within two years), the gain is not taxed.

(b) Gain attributable to the first $35,000 of sale price is excluded from income on sale of a personal residence by a taxpayer 65 or over.

A loss is not deductible, but may become so if the property is rented before sale.

A residence is a capital asset of the taxpayer which will ordinarily produce gain or loss when sold. Gain is capital gain and recognized unless the taxpayer reinvests the proceeds in another residence, or if he is 65 or over and elects to exclude the gain on the first $35,000 of sale price. Losses are personal nondeductible losses except for casualty losses. See § 4.16.

If part of the residence is used for business purposes, see § 3.19. If a farm is sold with the residence, the prorata share of sale price and basis will be carved out to determine the separate gain or loss on the residence. The special rules apply to a residence sold with the farm.

(a) Replacement of residence

One who sells or exchanges his residence at a gain is not taxed if he reinvests the proceeds in another residence and uses it within eighteen months or if he has already bought a new residence within eighteen months prior to the sale. Thus, if the new residence costs as much or more than the price received for the old, none of the gain is recognized for income tax purposes. If the new residence costs less, then the gain is recognized to the extent of the difference between the adjusted sale price (sale price minus fixing-up costs) of the old and the cost of the new.

If the taxpayer builds a new home, he must complete it within eighteen months before, or two years after, the sale of his old residence, and construction must have begun before the expiration of eighteen months from the date of the sale of the old residence.

The nonrecognition-of-gain provisions apply in case of a husband and wife who used both the old and new residences as their principal residences without regard to how they actually held the titles to the old and new residences. But the gain would be recognized if a son sold his house and took the new one in the name of his mother.

There is no way to get an extension of the eighteen months or two year periods for replacement of the residence without recognition of gain. Even a severe illness or a prohibition by a state agency which made it impossible for the taxpayer to buy a new house within the time period could not be considered because the statute allows no leeway.

Residence on Farm

The farmer who sells his farm with the residence must allocate part of the purchase price and the basis of the farm to the residence. Frequently an allocation of original cost was made when the farmer first made out a depreciation schedule for income tax purposes. Since depreciation claimed on farm business buildings was normally dependent upon what part of the total price of the farm was considered to be for the house at that time, it is ordinarily necessary to abide by the basis allocations then made. An appraisal or an allocation in the sale contract will help to establish the price received for the residence.

If a new farm is purchased, the farmer faces the dilemma of attributing value to the residence to avoid future gain or to the depreciable property to give himself a higher basis for long-term depreciation.

Amount of Gain Recognized

The gain on the sale of the old residence is the excess of the amount realized over the adjusted basis of the property. However, the gain to be recognized is only the excess of the "adjusted sale price" of the old residence over the cost of the new residence. The adjusted sale price is the amount realized on the sale less the "fixing-up" costs to assist in the sale of the property.

Where a purchase of a new residence results in the nonrecognition of a gain on the sale of the old, the gain not recognized must be subtracted from the cost of the new residence to determine the basis of the new residence. For example, if the new residence cost $50,000 and if $8000 of gain were not recognized, the basis of the new residence would be $42,000.

If the old residence was sold on an installment sale, the recognized gain would be spread over the period of installment payments. See § 3.26.

Condemnation or other involuntary conversion of a residence will fall under the involuntary conversion provisions unless an election is made to treat it as a sale and repurchase. In most cases it probably will make no difference. However, the involuntary conversion provisions may permit a longer period of time in which the property may be replaced, but the sale of residence provisions will cover the situation in which the replacement property was purchased prior to the condemnation, or threat thereof, of the old residence.

Losses on the Sale of Residences

Losses on the sale of personal residences are not deductible because they are personal losses. Sometimes, if the residence is rented for a period of time before sale, it is converted to business property, and the loss is allowed. The basis for determining loss is the cost or the fair market value at the time of conversion to business use, whichever is lower. The basis, of course, is adjusted for depreciation claimed and any other proper item of reduction or addition.

To claim a loss on the sale of a residence, it is imperative that the property be rented; even a ninety-day lease with an option to buy may suffice. However, if the property is held for sale, but not rented, the taxpayer is entitled to deduct depreciation and expenses for repairs, maintenance, insurance, and other expenses, inasmuch as putting the residence up for sale converts it to property held for the production of income.

(b) Exclusion of gain by person 65 or over

An individual aged 65 or over may elect to exclude from income the capital gain attributable to the first $35,000 of the sale price of his personal residence. If the sale price exceeds $35,000, the excluded gain is the same proportion of the total gain as $35,000 bears to the adjusted sales price.

The property must have been owned and used by the taxpayer as his principal residence for at least five years out of the eight-year period before the sale. The farmer must be 65 before the sale.

Amount of Gain Excluded

If the "adjusted sales price" is $35,000 or less, all of the gain is excluded. The "adjusted sales price" is the selling price reduced by the selling expenses and the fixing-up expenses.

> **Example:** A residence owned for ten years was sold by an over-65 taxpayer for $38,000. The property had a basis of $20,000. The commission on the sale was $1900 and the taxpayer spent $1500 to fix up the house for sale. The excludable gain would be computed:

Selling price		$38,000
Less: Commission	$1,900	
Fixing-up expenses	1,500	3,400
Adjusted sales price		34,600
Less: Basis		20,000
Excluded gain		$14,600

If the adjusted sales price is over $35,000, the exclusion is limited to a portion of the gain. The excluded gain is the part of the gain that is in the same ratio to the total gain as $35,000 is to the adjusted sales price.

Example: A residence was sold for $54,000. The commission was $3000, the fixing-up expenses were $1000 and the basis was $30,000.

Selling price		$54,000
Less: Commissions	$3,000	
Fixing-up expenses	1,000	4,000
Adjusted sales price		50,000
Less: Basis		30,000
Total gain		20,000
Ratio of $35,000 to $50,000		70%
Amount of gain excluded		14,000
Gain recognized		6,000

Special Rules

If property is held as community property, tenancy by the entirety or joint tenancy between husband and wife, if a joint return is filed, and if one spouse satisfies the age, holding and use requirements, the sale will qualify. A surviving spouse who is 65 or older may claim the exclusion if the deceased spouse satisfied the holding and use requirements during the eight-year period prior to sale and made no election with respect to a prior sale.

If only a part of the property was used for a residence, the gain applicable to the residence portion qualifies for exclusion. On the sale of a farm, the residential portion must be carved out. A residence condemned, destroyed or otherwise involuntarily converted is treated as sold, and the gain may be excluded.

The replacement provisions relating to sale and repurchase of a residence and involuntary conversion apply even though part of the gain is excluded under this section. The remaining gain is postponed if another residence is purchased.

One-Time Election

The privilege of excluding gain on the sale of residence may be used only once during the lifetimes of the taxpayer and his spouse. A prior election made by a spouse while he or she was married to someone else bars an election with the current spouse.

Farm Income Tax Manual, § 334.

§ 3.29 INDIRECT AND NONCASH INCOME

Income in the power of a cash-basis taxpayer to control will be considered income in that year, regardless of whether it is actually received.

Income may result from a number of unlikely sources. Herein are some of the more important traps for the unwary.

When funds are credited to or available to a cash-basis taxpayer without restriction, he is deemed to have received them for income tax purposes. Thus, interest on bank deposits, matured interest coupons on bonds, and dividends on stock are income to a taxpayer regardless of whether he actually collects them. The test is whether he could have had the money without substantial restrictions. The fact that some effort is necessary, such as presenting the deposit book, detaching and mailing the coupon, etc., is not a substantial restriction.

Many farmers have the impression that they can defer income simply by holding a buyer's check until another year or by asking the buyer of their products to pay in the following year. This will seldom work.

A check is income when received, not when it is cashed. A check for hogs delivered which the farmer could have picked up on December 30 was available income even though he did not get to town until January 2.

In another case a taxpayer delivered wheat to an elevator in April. Always before he had been paid at the time of delivery of the wheat but at his request payment was held up until the next year. He constructively received the income in the year of sale. But, where a farmer had a regular practice of contracting to sell wheat in one year for delivery and payment in another, that bona fide contract governed the year of receipt.

Deferred-Sale Contracts

A farmer who desires payment next year for his grain or other products sold this year should have a definite written contract which precludes him from getting his money during the year if he changes his mind and wants it. The Internal Revenue Service has challenged deferred-payment contracts on the ground that the contract is assignable and therefore the farmer could convert the right to money into cash. If the contract is a bona fide, arm's-length sale, provides a small down payment, fixes the dates of subsequent payments and declares that it is nonassignable, the right to income does not arise until the contract date of payment. An arrangement with the elevator to store the grain until the year of sale would be an effective way to defer income.

Mere postponement of payment will not push income into a later year; it is essential that the right to income be postponed. A restriction instigated by the taxpayer solely for tax juggling is seldom enough to stave off constructive receipt. However, where there is an element, perhaps only a small element, of business reason or purpose, the doctrine generally does not apply.

The Government's reaction to postponing income was illustrated in Revenue Ruling 70-294. A farmer disposed of his cattle through a livestock market under a contract whereby the cattle would be sold at auction and the farmer would be paid half the sales proceeds in one year and half in the next. The Service treated the entire sales proceeds as received in the year of sale on the theory that the livestock market was the sales agent for the farmer and that the Packers and Stockyards Act prevented a sale to the livestock market or to its personnel under which there could be a deferred payment. The ruling on the livestock sale distinguished the sale of grain to an elevator because the elevator owners are not the agents of the farmer and, as independent buyers, they could contract to pay in the future.

Unlikely Sources of Income

If an employer sells an item to an employee at a bargain rate, the bargain is regarded as ordinary income to the employee from compensation. However, a purchase at wholesale prices or standard discounts has not been regarded as producing income to the buyer.

Waived commissions on insurance policies and real estate sales are income to the seller. For example, an insurance agent sold insurance to his friends, relatives and employees, waiving his commissions due on the policies. He was taxed on the waived commissions. This is true even if he sells a policy to himself. A real estate salesman suffered the same fate.

Generally the passage of value without donative intent will result in income. A businessman's gift to another businessman who had suggested some possible customers was income to the recipient. A professional football player who was "loaned" the use of a car by a dealer had income.

The rental value of a corporation-owned house occupied by the controlling shareholder was taxable income in the absence of circumstances justifying the living quarters as required by his employment. Small farm corporations fall into this trap when the residence is transferred to the corporation with the farm property. The value of the use of the company car and the club dues paid by the corporation may be treated as income to a shareholder.

The Commissioner challenged a landlord as receiving rental income resulting from the services rendered by a tenant in clearing 75 acres of land. He contended that the services represented income which should be regarded as rent or in lieu of rent. The Government lost that one in court, but the theory will be argued again.

Advance premiums, prepaid premiums, or premium deposit funds which result in a reduction of insurance premiums due produces income to the extent of the reduction in premium.

The farm produce consumed by the family on the table is additional income to the farmer unless he reduces his deductions by the cost of raising the home-consumed food. Similarly, groceries withdrawn from a grocery store by the grocer and his sons were added back into income. See § 4.15.

It is firm law that the debtor receives income when his creditor forgives a debt due. Only in rare cases can it be argued that the creditor really made a gift from motives of "detached and disinterested generosity," to steal a phrase from the United States Supreme Court.

A transfer of property to a spouse under a property settlement incorporated into a divorce decree results in recognition of taxable gain to the conveyor spouse. The wife received the property in full satisfaction of her marital rights and released the husband from any further obligation. The gain is measured by the difference between the basis of the property to the husband and its fair market value at the time of transfer. It is just the same as if he sold the property, recognized his gain, and turned over the money to the spouse. But if the price is paid in installments, a court has held that the unstated interest rule does not apply.

The same theory was applied in the case of a shareholder who transferred stock to a key employee to keep him in the business. The transfer was regarded as a sale of stock, and gain was recognized to the extent of the difference between the basis of the stock and its value at the time of the transfer.

A Service ruling stated the answer to an old problem. A tenant farmer who is entitled to occupy a dwelling situated on the farm does not receive income as a result of his occupancy of the dwelling.

Diversion of Income

The Government often tries to treat as the father's income the gains from the sale of livestock which he gave to his children and raised for them. Only if the records are clear and the children are properly charged with the expenses of raising will the income be treated as belonging to the children-donees.

The Government also challenges efforts to give away property and then lease it back so that deductible rent is paid to children or trusts and income is spread among the family members, but taxpayers have been able to resist most of these challenges where they have been careful about the form of the transaction.

In Revenue Ruling 69-102, the Service held that a father who gave an annuity policy to his son had to report the gain when the son surrendered the policy.

Farm Income Tax Manual, § 347.

chapter 4

FARM BUSINESS EXPENSES

Section

§ 4.1 INTRODUCTION

A farmer may deduct all of the ordinary and necessary expenses incurred in his farm business or in connection with the mainte-

nance of income-producing property. No deduction is allowed for illegal payments, personal expenses, or capital expenditures. The Tax Reform Act of 1976 imposes a number of new limitations on farm losses.

Expenses of a farmer fall mainly into three categories:

(a) Ordinary expenses of the trade or business discussed in this chapter.

(b) Expenses incurred in connection with property held for the production of income or in connection with the management, conservation or maintenance of income-producing property. These are often referred to in this chapter because they follow closely the rules for business expenses.

(c) Expenses that are purely personal. Some personal expenses are deductible because of a specific statutory provision; for example, medical expenses, charitable contributions, property taxes on residence, etc. These are not discussed except in so far as they may relate to farm business expenses. Some expenses that appear to be personal may have a business element so that a part of the expense may be deductible.

The Tax Reform Act of 1976 added a number of new limitations on the deduction of farm expenses.

1. After 1975, a farm loss is allowable only to the extent the taxpayer has an amount "at risk" in the farm operation. A taxpayer is "at risk" to the extent of his cash investment, the adjusted basis of other property contributed to the farm operation, borrowings used in the farm business for which the taxpayer has personal liability, and the net value of non-farm assets securing nonrecourse borrowings used in the farm business. Earnings invested in the farm increase the amount "at risk" while losses decrease it.

2. Farm syndicates (a) may deduct expenses for feed, seed, fertilizer, and other farm supplies only when consumed (not when paid), (b) must capitalize poultry costs, and (c) must capitalize the costs of planting and maintaining a grove or vineyard until it becomes productive. A farm syndicate is a highly technical term, but generally if the sale of interests in the farm business was subject to **any state** or **federal** securities regulation or if 35% of the losses are allocated to limited interests, the operation is a farm syndicate. A limited interest is one in which losses are limited (as in a limited partnership) or one where the investor does not participate in management. There are some special exceptions for what might be called "real" farmers and their families.

3. A farm corporation which after 1975 ever has $1,000,000 in gross receipts must use accrual accounting and capitalize pre-productive expenses unless it is a Subchapter S corporation or 50%-owned by one family.

These rules were aimed primarily at farm tax shelters, but they may reach some others.

Necessary and Ordinary Expenses

Subject to these rules necessary and ordinary expenses are deductible. "Necessary" does not mean indispensable. It is sufficient if there are evident business ends to be served. The concept of "ordinary" does not require that the expenditures be either habitual or normal in the sense that this farmer either makes or is required to make them often, but that they are typical, normal, or common to farmers generally.

An expense is usually deductible by the taxpayer who incurs the liability. Problems regularly arise with respect to taxpayers who apparently have paid the expenses of another and seek to deduct them. As a general principle, taxpayers may deduct only their own expenses (or expenses properly attributable to them) in the same way that they must report income earned by them or earned by property owned by them. Diversion of deductions is no less tax evasion than diversion of income.

A check is income to a cash-basis taxpayer when received, and a payment by check is deductible when made even though the check is cashed in the next year.

Illegal Payments

The law explicitly denies deduction for some illegal payments. Fines and penalties paid to a governmental unit are not deductible. Similarly, bribes and kickbacks paid to a governmental officer are not allowable. Also any bribe, kickback, or other payment illegal under federal law or which could result in a criminal penalty or loss of license under state law (if such state law is generally enforced) is not deductible. Referral fees and similar kickbacks paid by one receiving payments under the Medicare-Medicaid Program are not allowable. Presumably, payments generally in violation of public policy may be deductible as ordinary and necessary business expenses unless falling under one of these prohibitions.

Personal Expenses

Personal expenses are not deductible unless there is a specific statutory provision permitting the deduction. In all considerations of the deductions listed in this chapter or in considering the possibility of deductions in planning a farmer's course of action, every effort must be made to see the business aspect of a deduction that might appear personal upon preliminary observation. For example, the charging off as a business expense of part of the cost of automobile operation, part of the food costs attributable to a hired hand, and the business portion of the home telephone are examples of the fundamental principle that personal expenses are not deductible, but that if a business portion can be shown, that much is deductible as a business expense.

A well-prepared return will show on its face that only the "farm share" of deductions has been claimed and that costs of raising farm products consumed by the family have been eliminated or income has been added to compensate. See § 4.15.

A personal expense can masquerade as a business deduction. For example, a supplier sold goods at two prices, one charge reflected normal prices and the other charge was higher because he gave premiums such as vacation trips and household appliances. When the taxpayer paid the high price and used the premium for personal or family purposes, the amount of his business deduction was the normal price. The excess was a personal expense.

Premiums paid on life insurance policies which were obtained to secure the taxpayer's debts were not deductible. Although his creditors were named as beneficiaries, he would have benefited by the payments because the policies would have reverted to him on payment of the debts or upon his death the proceeds would have paid off the indebtedness for the benefit of his estate. The expense was personal.

Business Use of Residence

Before 1976, residence expenses were allocated between farm business use and personal use, and the amount attributed to the farm business was deductible. The 1976 Act imposed some severe new restrictions. Even an allocable share of expenses is not deductible unless they arise from the exclusive and regular use of a part of the residence as either the taxpayer's principal place of business or as a place to meet customers in the ordinary course of business. A separate structure regularly and exclusively used for business is excepted from these rules. Also, the exclusive use test need not be met if the taxpayer is a retailer and part of the residence is used as storage for inventory. An employee must show that his use is for his employer's convenience.

This formulation requires some adjustment to assure deductions for the farm office. To qualify, no other use can be made, and the use must be frequent enough to be "regular." Even then the deductions may not exceed the gross income from the property as reduced by deductions which are allowable even if there were no business use, e.g., taxes and interest which remain deductible as itemized deductions even if these new rules apply. There are no guidelines for ascertaining the income from the property when it is used regularly and exclusively.

A farmer's office in the residence should be regarded as his principal place of business and his income is the income from the farm. It will be important to set aside a room to be used "exclusively and regularly" as the farm office. Make it look like a business office and use it as a business office.

Deductions for so-called vacation homes are even more severely restricted. Except for deductions allowed other than as business or income-producing deductions (e.g., taxes and interest), no deduction is allowable if the taxpayer occupies the place more than the greater of fourteen days or 10% of the number of days the property is rented. If rented less than fifteen days, then no business deductions are allowable. If, under these rules, some amount is allowable, it must be allocated between the personal and business portion. Basis for determining depreciation is the value when business use began or the adjusted basis on that date, whichever is lower. That amount should then be reduced by salvage value and depreciated in the usual fashion.

Capital Expenditures

Another common problem involves the principle that any expense which produces a benefit or results in the acquisition of an asset that lasts for more than a year is not currently deductible but must be amortized or depreciated over its period of benefit or use. The problem arises particularly in respect to deductions for supplies and repairs. Each expenditure must be analyzed to determine whether it is properly "current" or not.

This has many refinements. For example, depreciation on machinery and expenditures for labor and supplies incurred by a lumber company in reforestation of a timber tract were nondeductible capital expenditures. These costs could be recovered only as a depletion deduction when the timber was ultimately sold. Cost of access roads to the timber were also capital expenditures.

A race horse trainer entered into a joint venture with another person who provided the entire purchase price of a horse. The trainer was to train and board the horse at his expense and all racing earnings were to be shared equally. The Commissioner contended that all of the expenses of boarding and training the horse were nondeductible as capital expenditures for the acquisition of the trainer's half interest in the horse, but the Tax Court held that half of the expenditures should be treated as capital and the other half as deductible business expenses.

A farmer owned stock in a mutual assessment corporation which furnished water to the shareholders. A dam owned by the corporation was washed out by a flood and it became necessary to construct a new one. To obtain a bank loan, the corporation had to buy stock in the bank. The corporation assessed its members for payments to the bank and operating expenses.

No deduction was allowed for the portion of the assessments used to repay the bank loan nor for the purchase price of the bank stock. Only the portion of the assessment which represented operating expenses was was deductible. The costs of the bank stock and the replacement dam were capital expenditures. The assessments could not be deducted as soil and water conservation expenses, nor could the farmer claim depreciation on the dam because it was owned by the corporation, nor could he claim a casualty loss deduction for the washout of the old dam.

Attorney's fees paid in securing the reduction of a special assessment imposed for local benefit were not a deductible business expense, but a capital expenditure added to the basis of the property.

Lease or Purchase

Many farmers lease assets used in the farm business and deduct the rental payments. If the transaction is really a lease, the deduction is proper. But the Internal Revenue Service thinks that some leases look very much like conditional sales contracts under which the farmer is buying the asset. If he is acquiring an asset, the transaction is a sale and he deducts not rent but depreciation and interest.

In determining whether the lease is more like a sale, the Service looks to (1) whether the rents and the option price to acquire the property at the end of the lease roughly equal the purchase price plus interest; (2) whether high rents in the early years of the lease approximate the purchase price and the farmer retains the use of the asset in later years at a relatively low rental; (3) whether the farmer will acquire title at the end of the lease for a nominal price and the rental payments are credited as a substantial part of the purchase price; or (4) whether on the face of the whole transaction the farmer obtains the use of the asset or the ownership of it by payments which do not conform to the reasonable rental value of the asset.

Careful consideration of the recovery of cost of machinery through 20% depreciation and fast methods of depreciation plus the deduction for interest or carrying charges may make the lease arrangement less attractive.

Political Contributions

An individual who makes a contribution to or for a political candidate may claim either:

1. a credit against his tax for an amount equal to one-half of his contributions up to a maximum credit of $25 ($50 on a joint return), or
2. an itemized deduction for the full amount of the contribution up to a maximum deduction of $100 ($200 on a joint return).

Contributions to political candidates or to political committees qualify for a credit or deduction. Contributions to political candidates qualify only if they are made for announced candidates for use in connection with a candidacy and are verified. Contributions to a national, state or local committee of a national political party are eligible for the credit or deduction without reference to the political candidate involved.

Only individuals may claim the deduction. The credit is granted by Section 41 and the deduction by Section 218. These are mutually exclusive sections; the taxpayer must elect one or the other. Estates, trusts and corporations may not claim either.

For a contribution to a political candidate to qualify, the candidacy must be announced by the end of the year. Expenses of a candidate which are political in nature qualify if they are incurred during a calendar year in which the candidate is an announced candidate. Expenses incurred after the campaign qualify only to the extent that they are directly related to the nomination or election. The contribution may be made before announcement of the candidacy.

Generally, a qualifying contribution may be made to:

—any candidate for nomination or election at the Federal, state or local level running in a primary, general or special election;

—any committee, association or organization organized to seek nomination or election of one or more candidates and which so uses its funds;

—national, state or local committee of a national political party.

Farm Income Tax Manual, §§ 400, 714.

§ 4.2 FARM BUSINESS EXPENSES GENERALLY

Most farm business expenses are routine, easily recognized as proper deductions, and cause little or no problem. Several, however, warrant comment:

(a) Machine hire

(b) Gasoline and other fuel

(c) Insurance

(d) Rent of farm or pasture

(e) Freight and trucking

(f) **Dog and license**

(g) **Baby chicks and custom hatching**

(h) **Attorney and other professional fees**

(i) **Lobbying expenses**

(a) Machine hire

Machine hire for moving earth, building ponds, etc., is a deductible soil and water conservation expense or land-clearing expense if the election is made. If not, no deduction is allowed and the cost is added to the basis of the farm. See §§ 4.13, 4.14.

The cost of equipment and labor hired to build fences, bins, feeding platforms, additions to structures, etc., is not deductible but is included as part of the depreciable cost.

(b) Gasoline and other fuel

The deduction for gasoline and other fuel must be reduced by any co-operative dividend or gas tax refund credit claimed, unless such items are treated as income. The gasoline tax refund is excluded from the deduction on the accrual basis, because the refund accrues as the gasoline is used. The deduction is the cost of the gasoline less tax. A cash-basis farmer should deduct the full cost and include the refund credit as income. See § 6.6.

Automobile gasoline expense is deductible only to the extent that the car is used for business. It is usually more convenient to segregate all automobile expenses and compute the percentage attributable to business use.

The cost of fuel for heating, lighting, or cooking in the farmer's home is not deductible except to the extent that these expenses are incurred for hired help or other business purposes.

(c) Insurance

Premiums on fire, windstorm, theft, or any other type of casualty insurance on assets used in the business, premiums on life insurance of employees paid as additional wages, and premiums paid on health and accident insurance policies for employees are deductible. A self-employed farmer may deduct premiums on insurance to pay overhead expenses while disabled. Amounts set aside as a reserve for self-insurance are not deductible.

Premiums on three- and five-year policies of fire or casualty insurance must be prorated over the life of the policy even though the taxpayer is on the cash basis.

Premiums on insurance covering the dwelling house or any other asset not used for business purposes are not deductible. Nor are premiums on life insurance on the farm owner's life deductible.

Automobile insurance is usually more conveniently deducted on an automobile expense schedule where the percentage attributable to business use of the car can be computed.

(d) Rent of farm or pasture

The value of crops delivered by the tenant to the landlord is not deductible but all costs of raising the crops are.

No "rent" reduction is allowed if the taxpayer is acquiring or will acquire an equity in the property. He is then a purchaser. See § 4.1. See § 4.6 for the difference between rent and pasturage.

(e) Freight and trucking

Expenses for hauling personal items such as fuel for heating the home are not deductible.

Trucking expenses on materials used to build new buildings or make improvements on old buildings are part of the depreciable cost of the new asset or improvement, but transportation costs on materials used for repairs are deductible as freight or repairs.

Transportation expenses in delivering livestock to the farm should be capitalized and added to the cost of the animal. Transportation to market is an expense of sale.

(f) Dog and license

The cost of licenses for dogs kept for business reasons (killing rats, herding sheep, etc.) is deductible. Although the cost of a business dog should be a capital expenditure recoverable by depreciation, if the cost is small it might, for accounting convenience, be deducted in the year of purchase.

(g) Baby chicks and custom hatching

The purchase price of baby chicks, ducks, turkeys, etc., or custom hatching costs are deductible when paid if the practice is followed consistently. Costs of larger livestock must be offset against the sale price in the year of sale.

(h) Attorney and other professional fees

Fees of an attorney, accountant or other consultant for services in connection with farm business matters, including any expense incurred in connection with the determination, collection or refund of any tax are deductible if not a capital expenditure. This includes the portion of the estate planning fee allocated to tax advice.

Generally, expenses for the acquisition, perfection or recovery of title to property are capital expenditures, not currently deductible, but if they relate to the operation of the business or the continued use of property for income-producing purposes they may be deducted.

(i) Lobbying expenses

Deduction is allowed for expenses incurred in limited forms of more or less "respectable" lobbying in direct connection with (a) appearances before, or statements to, a legislative body with respect to legislation or proposed legislation, or (b) communication of information between the taxpayer and an organization of which he is a member with respect to legislation or proposed legislation in which he and the organization are interested. In addition, the part of the organization dues attributable to expenses of such activities is deductible.

Except as provided under § 4.1, no expenses are deductible for activities in any political campaign on behalf of a candidate for public office or for efforts to influence the public on legislative matters, elections or referendums.

Farm Income Tax Manual, §§ 404, 406, 408, 411, 415 to 417, 424.

§ 4.3 EXPENSES OF HIRED LABOR

All expenses paid for hired labor, including wages, bonuses, cost of board and lodging and payments to move a new employee, are deductible. Children may be employed by parents. In some cases meals and lodging and moving expenses may not be income to employees.

Wages paid to a hired man and to occasional help are, of course, deductible. Cost of meals and lodging furnished hired help is also deductible. If an accurate per person estimate can be made, it should be used. Otherwise, the best estimate of cost, not value, in the community is satisfactory.

Value of products raised on the farm and used for the board of hired help is not deductible. Only purchased items may be deducted. The portion of the expense allocable to the family must be excluded.

Bonuses to hired help are deductible as wages and are subject to Social Security withholding and, where appropriate, income tax withholding and unemployment taxes. Farm employers are not usually concerned with the latter two. But, see below for new rules on unemployment insurance.

Value of a turkey, ham or other items of merchandise of nominal value distributed at Christmas or other holidays to employees as a means of promoting good will are deductible as business expenses but are not taxable income to the employees.

Cash, gift certificates, or similar items of readily convertible cash value, regardless of the amount involved, are deductible either as wages or business expenses and are income to the employees.

Wages and other costs such as meals and lodging of a maid or housekeeper, to the extent that her services are rendered to hired help or otherwise contribute directly to the farm business, are deductible.

Any other expenses incurred because of hired help are deductible, for example, premiums paid on health or accident insurance policies of employees, repairs on the dwelling to the extent occupied by hired help, and depreciation on bedding, furniture and part of the dwelling to the extent occupied by hired help. The value of ten Jersey cattle butchered and given to employees as bonuses was additional compensation and therefore deductible.

Amounts paid for labor on new buildings or improvements on old buildings are capital investments added to the depreciable cost of the improvements. If the labor is for repair purposes, it is deducted as a repair if the repair job is contracted for, but as labor if it is done by the farmer's employee.

The value of the farmer's own labor or that of his wife or dependent children is never deductible. Deductible wages may be paid to children or to a spouse. In the latter case, a joint return (which is usually very favorable) will cancel the benefit of the deduction.

Wages to Children

Wages in good faith paid to children for the reasonable value of services rendered may be deducted. Children under nineteen years of age, and those over nineteen if attending an accredited educational institution or receiving full-time educational on-the-farm training, may be dependents even though they earn more than the amount of the dependency exemption.

A child using the low income allowance plus his exemption of $750 may earn $2450 free of tax. The parent may still take the regular exemption of $750 for the child if he is under nineteen or attending school. The cost of board purchased for working children should be deducted, but not meals and lodging furnished to an unemancipated child.

Payment to Settle New Hired Hand

An employee may deduct moving expenses (including meals and lodging in transit) for himself and his family incurred to start work in a new location. He must move at least thirty five miles (fifty before 1977) and he must work in the new area for at least 39 weeks. If the employer reimburses the employee, the employer deducts the payment and the employee reports the reimbursement as income and claims his deductions.

Deductible expenses also include travel costs in searching for a new residence, meals and lodging while occupying temporary quarters in the new location for up to thirty days, and the expenses of selling the old house and purchasing a new one or settling an old lease and acquiring a new one. These expenses are limited to $3000 ($2500 before 1977) of which only $1500 ($1000 before 1977) can be for house-hunting trips and temporary living costs.

Self-employed persons are now permitted the moving expense deduction if they work in the new area for 78 weeks. Thus, a tenant would be entitled to the deductions if he meets the qualifying tests. If the landlord paid for the move, the tenant would report the payment as income and claim the offsetting deductions.

The deduction is taken in the current year even though the time test is not yet met. The deduction is taken from gross income; it is not an itemized deduction.

Meals and Lodging

Meals and lodging furnished by the employer may be excluded from income by the employee if (1) meals are furnished on the business premises of the employer for the employer's convenience, and (2) lodging is furnished on the employer's premises for the convenience of the employer and the employee is required to accept such lodging as a condition of his employment. For example, a farm worker receives his room and board on the farm primarily because he is expected to be on hand for early morning chores, late evening chores and emergency duties.

Rental value of living quarters furnished on the business premises by the employer to the employee is excludable from the employee's gross income even though the value of the living quarters is considered a part of the employee's compensation and is deducted from his salary. If the employer furnishes utilities, the cost is treated as nontaxable lodging, but if the employee buys his utilities from the suppliers, he cannot deduct the cost.

Even though owners or partners do the same work as employees, the Government claims that there is no deduction for the cost of food and lodging furnished by the owners or partners to themselves under any circumstances. One case has held that the cost of food and lodging furnished to a ranch partner was deductible, so the issue is not closed.

An employee of a farm corporation (even though a stockholder) has an employer whose convenience may be served. For example, two shareholder-employees of a Subchapter S corporation which owned two Wyoming cattle ranches had no income from the food and lodging provided by the corporation and the expenses were deductible by the corporation. The "convenience of the employer" rule applied.

The Government and the courts take a dim view of the farm shareholder-employee who tries to get free living on the farm. They take the position that the lodging on the farm must not only be required (which the sympathetic Board of Directors can arrange) but also "necessary" to the adequate performance of the employee's duties.

The circumstances must be real. In one case, the board of directors of a poultry-breeding corporation passed a resolution making it a condition of the employment of its president and sole stockholder that he live in the residence on the poultry farm. In fact, the corporation rented the residence from the president and sole stockholder for $444 a month and then required that he live in it to keep an eye on the poultry operation. In prior years, he had done pretty well managing the poultry farm from a house in town. The food and lodging furnished to the taxpayer were not really for the convenience of the corporate employer and the president was required to include the fair market value of the meals and lodging in gross income.

Unemployment Insurance

Farmers will find a new expense in unemployment insurance. To be liable, a farmer must have ten or more employees for twenty or more weeks throughout the year or have a payroll of $20,000 in one calendar quarter. If $1000 or more is paid to domestic workers in any one quarter, there will be liability for unemployment insurance on those wages. This starts in 1978.

Farm Income Tax Manual, § 401.

§ 4.4 REPAIRS AND MAINTENANCE

A farmer's repair deduction properly includes:

(1) Building repairs—paint (on old buildings), shingles, siding, sand, cement, labor costs, etc.

(2) Fence repairs—posts, wire, staples, labor, electric fence parts, etc.

(3) Machinery and equipment repairs—tractor, cultivator, binder, pumps, harness, sharpening plows, blacksmithing, etc.

(4) Miscellaneous repairs—water tank, feed platform, etc.

(5) Small tools purchased.

A taxpayer now has a choice about the deduction of repairs. He may fight it out along the lines suggested in this section or may choose to use a "repair allowance" option. This election is an annual one and may obviously be manipulated to the taxpayer's advantage.

Capital expenditures for labor or material to construct buildings or overhaul machinery, or for improvements or rebuilding which add materially to the value or prolong the life of the asset are not repairs but are added to depreciable cost. The complete renovation of property in the year of purchase is similarly treated.

Generally items incidental to an over-all plan of rehabilitation, improvement or modernization must be capitalized. Alone, each item might be a repair but all together they will be treated as a capital expenditure. The same is true of painting a tenant house and other buildings just before selling. Painting and papering are ordinarily deductible repairs, but not if the property is sold in the same year.

Expenditures are not currently deductible if they tend to prolong the life of the property or are "one shot" affairs that probably will not be repeated, even though there is little or no increase in the value of the property as a result of the expenditure.

The line between repairs and capital expenditures is always difficult. To justify the immediate deduction, the farmer must strive to find that the cost maintains the status quo and adds no additional value.

Auto repairs should not appear in the farm repairs schedule; such repairs are deductible only to the extent of business use of the car. It is clearer and aids computation to segregate all automobile expenses as one item.

Cost of labor or material for repairs to the residence may not be deducted because it is not an asset used in the business. But such repairs are deductible to the extent the home is occupied by hired help or otherwise used for business purposes.

Even though the Treasury's "Tax Guide for Small Business" says that you can deduct "the cost of repairs, including labor . . . ," it does not mean the value of your own "do-it-yourself" labor.

Explain Repairs

An explanation of the deduction claimed by itemizing repair costs on another part of the return or in an attached schedule is desirable to show that large repair totals were carefully computed and are composed of many small items difficult to challenge.

Items should not be in the repair schedule if they can properly be classified under another deduction. Repair expense is probably the most abused deduction on the farm return. Auditors will look at a high total with suspicion justified by the frequency with which farmers include material and labor expenses on improvements as "repairs." Consequently, it is well to exclude from this classification all items which may reasonably be entered elsewhere.

Small tools of short life need not be depreciated but the expense may be charged off in the year of purchase. As a practical matter, tools are often added to supplies (§ 4.8) or listed as a separate item, "small tools." This minimizes the chance of an audit for the purpose of determining whether capital expenditures were included in repairs.

Repair Deduction Allowance

The depreciation regulations now contain an election which will avoid the repair versus improvements argument. The rules are complex, but generally speaking a taxpayer has an election to make with respect to expenditures for repair, maintenance, rehabilitation, or improvement of property. He can choose to treat them under the regular rules for determining whether to capitalize or to deduct, or he can elect to deduct them up to an amount determined by applying a **repair allowance percentage** to the adjusted basis of all property in a guideline class of assets. Expenditures in excess of this amount are to be capitalized. If the election is made, the taxpayer must treat all such expenditures under these rules. The election is available each year and is made on the tax return for the year. It can be made for each guideline class of assets.

The kinds of property which qualify, the limitations on the use of the election, the expenditures excluded from the election, and many other detailed rules are all in the Regulations. Regs. § 1.167(a)-11(d)(2).

Farm Income Tax Manual, § 402.

§ 4.5 FARM INTEREST

Only interest that relates to the farm business or to farm business property should be deducted on the farm schedule.

Every contract of purchase or loan should specify the precise rate of interest.

All interest paid or accrued is deductible. See § 3.26.

Interest in excess of the amount the agreement provides is not deductible; for example, where the interest rate was fixed at 4% but the borrower paid 5%, the extra 1% was not deductible.

An accrual-basis taxpayer accrues the interest due, whether paid in advance or not. However, where a cash-basis taxpayer pays the interest in advance through a discount from the face or principal of the loan and the loan is payable in installments, he deducts a prorata portion on each repayment. The accrual-basis taxpayer deducts the amount actually accrued each year.

Interest paid on tax deficiencies arising out of a revaluation of livestock inventories or other business income items is properly deductible as a business expense. But a penalty is not deductible either as interest or taxes.

A penalty for prepayment of a mortgage is deductible as interest, but if a new loan is acquired, the penalty is amortized over the life of the loan.

Points, loan placement fees, and premium charges paid for the use of money by a borrower are interest. If the payments are charges for services by the lender, as the Service treats "points" under a VA or FHA mortgage, the payment is not interest. If the seller pays a loan charge for the borrower, it is an additional expense of sale.

Interest paid on an automobile loan is more conveniently allocated to business and personal use if included in "automobile expenses."

If state law permits a farm owner to elect to pay special assessments against real estate for public improvements in annual installments over a period of years at interest, the amount representing interest is deductible if it can be separately determined.

Interest charged on credit card purchases is deductible.

Interest on loans to acquire investment property is subject to a deduction limit.

Before 1976, this limit was the total of (1) $25,000, (2) net investment income, and (3) net long-term capital gains. The deduction for one-half of capital gains was lost to the extent that the deduction for investment interest exceeded the total of items (1) and (2) and was measured by (3). For 1976 and later years, the limit is $10,000 plus investment income. Interest on indebtedness to acquire interests in a family business may be as much as an additional $15,000 and be deducted. Any disallowed interest may be carried over for future deduction subject to these limitations.

Prepaid Interest

Before 1976, the Internal Revenue Service limited the deduction for prepaid interest to the amount attributable to the current taxable year and one additional year. But, the IRS still objected if there was a "material distortion of income." Larger prepayments were amortized over the period to which the interest related. In 1976 and thereafter, the only prepaid interest which may be deducted when paid is "points" on a home mortgage. All other interest paid in advance is deducted in the year in which the interest would ordinarily be charged.

Construction Period Interest

Interest during a construction period is subject to the same rules as taxes during that period. See § 4.9.

Carrying Charges

If a contract for the purchase of personal property has a "carrying charge" but no stated interest, 6% of the average unpaid balance may be deducted as interest. "Average unpaid balance" means the sum of the unpaid balances outstanding on the first day of each month during the year, divided by twelve.

Example: A farmer on June 21 bought a tractor. He was allowed a $500 trade-in on his old tractor leaving a balance of $2000 due including a "carrying charge" of $125. He is to pay $100 a month until the contract is paid up. The interest deduction would be computed as follows:

Unpaid Balances Outstanding:	
July 1	$2,000.00
Aug. 1	1,900.00
Sept. 1	1,800.00
Oct. 1	1,700.00
Nov. 1	1,600.00
Dec. 1	1,500.00
Total	$10,500.00
Average unpaid balance ($10,500 divided by 12)	$ 875.00
Interest deduction (6% of $875)	$ 52.50

The same result is reached (1) by multiplying the sum of the monthly unpaid balances by half of 1%, or (2) by taking the sum of the monthly unpaid balances, moving the decimal point two places to the left and dividing by two.

Applying this to the example above:

(1) Multiply $10,500.00 by .005. The result is $52.50.

(2) Move decimal point two places to the left ($105.00) and divide by two ($52.50).

The purchase contract must state a carrying charge to support a deduction, even under the 6% rule.

Under some charge-account arrangements, the customer signs an agreement under which he is to pay each month either the balance due or a specified portion of the balance due on his account. If he pays less than the balance due, a service charge on the unpaid balance is added to his account. If the actual interest charge cannot be ascertained, then the 6% rule will apply, but such amount cannot be in excess of the actual service charges made. Amounts paid under bank charge credit cards are fully deductible as interest.

Farm Income Tax Manual, § 403.

§ 4.6 FEED

Ordinary purchases of feed are deductible. So is the basis of crops which are received from a decedent and fed to animals. Advance deposits are not current deductions.

A cash-basis taxpayer ordinarily deducts his feed costs when he makes the payment. While an accrual-basis taxpayer usually inventories feed, a farmer using the unit livestock method need not inventory feed raised for feeding to livestock. The same rule should apply to other inventory methods, but may not.

If crops inherited from a decedent are fed to livestock by the estate or other successor, there is a feed-purchased deduction in the amount of the basis of the crops in the hands of the estate or other successor. If the crops are sold, the gain or loss is the difference between the selling price and the basis to the estate. See § 5.8 for determination of the basis at death before 1977 and after 1976.

If rents received as crop shares are fed to livestock, the Service holds that the value of the crop is treated as income and that there is an offsetting deduction for feed in the same amount. See § 3.4.

The cost of feed for draft, breeding, dairy, and sporting animals in transit to the farmer should be treated as part of the purchase price or capitalized cost of the livestock to be recovered on sale or through depreciation.

Advance Payments

If an advance purchase of feed is simply a deposit against future needs, it is not deductible until the expense is actually incurred. For example, a purported purchase of $23,000 worth of feed from an elevator to be delivered in the future as needed by a partnership in its feeding operations was treated as an advance deposit for future sales of feed. The court held that the acquisition of an asset which would last substantially beyond the close of the taxable year was not a current deduction. In fact, the $23,000 was three times the amount of any prior single payment for feed and in all prior ventures the partnership had paid for feed on or after delivery, not in advance.

The Tax Court denied a deduction to a cash-basis, calendar-year farmer who paid on December 31 for feed to be delivered over the following year. The quantities and prices were set out on the sales ticket but the feed dealer did not set aside any feed to meet the order. The farmer could substitute other feed or nonfeed products for the quantities presumably ordered. There was no charge for storage nor was the risk of loss shifted to the buyer. He got no special price or treatment for his advance payment. The Eighth Circuit Court of Appeals found that the contract was binding (not a deposit) and assured a supply of feed. The payment was deductible.

Generally, if the payment is unconditional and cannot be refunded, or the payment is necessary to achieve preferential treatment in the face of a pending shortage or if there is some other rational reason for the expenditure other than reducing current income taxes, the deduction should be proper. See the discussion at § 2.2-2.

The Service has published a stiff three-pronged test for deductibility of prepaid expenses. It provides that cash-basis taxpayers can deduct advance purchase payments for feed and other supplies in the year paid if the expenditure:

(1) Is a purchase and not just a deposit. If there is a binding commitment to take delivery of a fixed quantity at a fixed price and the buyer is not entitled to a refund or repurchase, the payment is not a deposit. Nor is it a deposit just because the deal allows a substitution of feed ingredients to meet changing diet requirements, nor must the material be delivered by year-end.

(2) Is for some valid business purpose, such as to fix a maximum price or to assure a feed supply or to gain preferential treatment in the face of a feed shortage.

(3) Does not materially distort income for the year. IRS will look to past practices of the taxpayer, the amount of the payment in relation to past purchases, the time of the year, and the size of the expenditure in relation to income. (IRS persists on this point though the courts have not yet passed on it.)

It is clear that the IRS is still on the offensive where advance payments are concerned. The firmer the contract terms, the greater the chance of satisfying the Service, and if, need be, the courts.

Here are some further suggestions which could strengthen the argument with the agent in the event a return is audited.

The prepaid purchase should be for a reasonable amount of feed for the livestock owned or expected to be owned in the following year; the buyer should pay all property taxes, insurance and storage fees on the feed, assume all the risks and burdens of ownership; and the prepayment must involve an actual cash outlay, not a note. The seller should separate all service charges such as yardage, mixing, handling, doctoring, etc., and he must treat the transaction as a sale on his books.

It would seem that these suggestions, as well as the IRS tests, should apply to fertilizer, chemicals, fuel, or other supplies that might be purchased late in the year for delivery and use next year.

Farm syndicates cannot get a deduction for feed until it is fed. See § 4.1. Obviously, an accrual-basis taxpayer must inventory the amount of supplies on hand and lose the deduction through the inventory. As noted above, a few farm corporations and partnerships may have to use the accrual basis. See § 4.1.

Pasturage or Forage

A payment for pasturage or forage—standing crops, legumes or grass left in a field—is properly a feed deduction, not an advance rent payment which would be deducted pro rata over the period of the lease. Generally the line should fall between existing forage at the time of acquisition and a feed crop or grass to be grown thereafter.

In most of the cattle country, the term "pasturage" is synonomous with "forage" and refers to a standing crop of grain, corn stalks, legumes or grass left in a field as a form of unharvested feed. As a practical matter, pasturage is found in areas of some degree of aridity as heavy rainfall and snow will cause excessive deterioration. The Mississippi River represents a general boundary which carries over into word usage.

East of the Mississippi, the phrase used is "renting a pasture;" west of it, the expression is "buying grass." In country in which it is dry enough for grass to "winter cure," the clump which is big enough for the cattle to eat at a specific moment has been growing for a year or two. If in a given summer the field is overgrazed, there will not only be no "feed," "forage" or "pasturage" for the coming winter and spring, but it might take a full year of non-use for the area to recover to normal carrying capacity.

When corn is harvested by mechanical pickers, three or four bushels per acre are left in the field and the stalks, which have a certain amount of nutrition and roughage value, are left standing. The common phrase for acquiring this "pasturage" is "buying a stalk field." This does not mean acquiring title to the land but only a right to the standing crop of forage left on it from a previous growing season.

The phrase "rent a corn field" means the acquisition of the right to use a bare field to plant a new crop. The termination of the land rental does not extinguish the right of the tenant to harvest the grain or feed his cattle on the stalks left standing in the field.

With alfalfa and clover, it is common to leave the "last cutting" standing in the field to provide "winter pasturage." If it is reasonably dry, this should be almost identical in feed value to hay that is actually put up.

When real estate is sold and the buyer acquires the "land and all on it," the Service has permitted a feed deduction for the reasonable value of any standing crops, forage or pasturage acquired with the land.

Where a bare field is acquired in November and planted to a grain crop or even an annual pasture grass such as sudan, the payment for the land is rent. In most agricultural areas, the IRS has permitted the rent to be deducted in the year of payment; however, the Service could properly require the payment to be amortized over the rental period.

If title has been acquired to a standing crop of corn, corn stalks, alfalfa or fescue grass, particularly in the fall of the year, and permission obtained to bring in cattle to consume it, the transaction is a sale of feed.

Where the acquisition of the land use for a year involves title to any feed existing on it at the time of acquisition plus any that might grow on it after the date of acquisition, there are elements of both pasturage and rent. The allocation will vary by crop and area. The agricultural university in a particular state should be helpful in making the breakdown.

Farm Income Tax Manual, § 405.

§ 4.7 FERTILIZERS AND LIME

Fertilizer and lime costs are usually current expenses, but if the value extends over several years, the expenditure should be prorated unless a special election is made.

Most fertilizer applications are regarded as current expenses, but lime and like materials often have a more lasting benefit. Where the piece of land limed is relatively small, the benefit is of short duration, and the farmer is likely to have similar expenses every year, a consistent practice of deducting the cost of liming in the year paid has not been questioned. There is evidence that even on large operations, such as trustees holding substantial farms, no issue has been raised on a consistent practice of deduction of liming costs.

In theory, the cost of fertilizer and lime should be spread over the period of usefulness. However, the cost need not be spread evenly if the benefits are greater in the early years. If the cost is to be capitalized, a declining allocation over several years is entirely proper.

However, a special statutory provision permits a taxpayer engaged in farming to elect to deduct expenditures for fertilizer, lime and ground limestone, marl or other materials to enrich, neutralize or condition land used in farming and the cost of application. The election may be made each year on the tax return by claiming the deduction.

"Land used in farming" and the "business of farming" have the same meanings as under the soil and water conservation deduction. See § 4.13. Expenditures on new land must be capitalized. See §4.12. A landlord whose rent is not a share of production is not in the business of farming.

Farm Income Tax Manual, § 407.

§ 4.8 SUPPLIES PURCHASED

Supplies are often purchased for cash, and almost without exception the small farm operator will neglect to record some of these purchases. Spotting omitted items will not only reduce the tax bill, but improve bookkeeping in future years.

Supplies include the cost of any item which was bought for use in the business of farming, but not the cost of work clothes if they are adaptable to general wear.

The cost of baby chicks and egg-laying hens may be deducted in the year of purchase if done consistently. To shift to this practice is a change in accounting method which requires permission of the Commissioner of Internal Revenue. See § 6.1.

Doubts as to whether an item is supply or a repair may be resolved in favor of supply. An item which would otherwise be a repair is properly treated as a supply if the farmer buys a number of the items at one time in order to have them available when needed.

Small capital items are often charged off as expenses in the year purchased even though the cost seems too high or the useful life too long for a supply item, simply because the amounts are too small to enter on the depreciation schedule. Ordinarily an item having a useful life of more than a year should be depreciated and the original cost should not be deducted in the year of purchase. However, a consistent practice of deducting items of relatively small cost which are in the nature of small tools (in that some are bought each year and no depreciation is claimed on similar articles previously purchased) is authorized so long as income is clearly reflected.

The size of the farm operation will have much to do with drawing the line between supply expense and depreciation schedule. A large farmer's supply item might go unchallenged while the same item on a smaller operator's return should be a depreciable item. Sometimes handy rules of thumb can be based upon dollar values; for example, items costing less than $100 or $200 might be treated as supplies for the average farmer in a particular locality. But there are cases in which the Service has challenged this practice.

Farm Income Tax Manual, § 409.

§ 4.9 TAXES

Most taxes incurred in the farm business are deductible as business expenses even if not as taxes. Some taxes are deductible only as personal expenses and some not at all.

Taxes incurred in farm operations are usually trade or business expenses, thus not deducted as "taxes." Some taxes are not deductible either as taxes or expenses: (1) federal income tax; (2) employee's or self-employed person's Social Security taxes; (3) gift or death taxes, and (4) taxes on real property treated as apportioned to another taxpayer.

If the tax is not a business expense, a deduction for taxes may be claimed for: (1) real and personal property taxes; (2) state, foreign and local income taxes; (3) state and local sales and use taxes; and (4) state and local gasoline taxes. No cigarette, liquor, amusement or similar taxes are deductible except as business expenses. The part of an automobile registration fee that is based upon the value of the car is deductible as a personal property tax.

Taxes not assessed in the name of the taxpayer are generally not deductible. Only the one on whom the tax is imposed can deduct a tax.

Taxes paid by a cash-basis taxpayer, if not merely a deposit, are deductible even though they are contested. The amount actually paid is deducted and a refund treated as income. An accrual-basis taxpayer who contests taxes is also allowed a deduction when he pays the tax and he, also, includes a refund in income.

Construction Period Taxes

Taxes on unimproved real property may be capitalized or deducted until a construction is commenced on it. Once construction is commenced, real property taxes must be capitalized until the development or construction is completed. Beginning in 1976, the amount capitalized may be amortized over a period beginning with four years for amounts capitalized in 1976 and being extended one year each tax year until a ten-year period is reached for taxes capitalized in 1982 and later years. In 1976, 50% may be deducted if the taxpayer chooses, and 50% is subject to amortization in the next three years. Instead of this amortization such capitalized taxes may be added to the basis of the property and depreciated with the property. Taxes on unproductive personal property may be capitalized or deducted up to the time of use. Property, sales, and use taxes are included. Other carrying charges may also be capitalized before use of the property begins or they may be deducted. Interest on real estate loans during the construction period are treated like real property taxes and must be capitalized. It can then be amortized or depreciated as part of the property cost.

Real Estate Taxes

Real estate taxes on farm property are deductible on Schedule F but the part of the real estate tax which is attributable to the dwelling house should be subtracted and entered as a personal deduction on Schedule A if deductions are itemized. The full real estate tax is, of course, deductible on Schedule F if all of the amount attributable to the dwelling is included in a homestead tax exemption.

Real estate taxes for the year of sale must be apportioned between the seller and buyer in proportion to the number of days during the real property tax year that each held the property. It does not matter who actually pays the tax. A cash-basis seller can deduct his proportionate part even though the buyer paid all the tax.

If one of the parties is liable for the tax at the time of sale, he deducts his share when he pays or accrues the tax. The other party is treated as having paid or accrued the tax on the date of sale. It is entirely possible that the parties will take the deduction for their respective shares in different years.

If neither party is liable for the taxes at the time of sale, the party holding the property at the time the tax becomes a lien is considered liable for the taxes. For example, if the buyer held the property when the lien attached, the other party, the cash-basis seller, would treat his share of the taxes for the year of sale as paid in the year of sale, while the cash-basis buyer would treat his share of the taxes as paid when he pays the taxes, which might be the year after the sale.

The tax sale of property by the county is not a payment of taxes by the owner of the real estate and no deduction is allowed.

Personal Property Taxes

Personal property taxes on household goods should be separated from the total personal property taxes paid. Taxes on business property are deducted on the farm schedule. The amount levied on household and other nonbusiness items is claimed as a personal deduction on Schedule A.

Auto license or property tax on the business auto should be included as "automobile expenses" where the percentage attributable to business use may be computed. If any part of an automobile license or registration fee is based upon value, that part is deductible as a personal property tax.

Sales and Use Taxes

Sales taxes on deductible items purchased will be deducted as part of the cost. If the purchased item is depreciable, the tax may be deducted or added to the depreciable cost of the purchased item. Usually sales taxes will not be deducted as taxes but will become a part of the cost of the item acquired. If the tax is not paid in connection with the trade or business, then it is to be deducted as a tax. Use taxes are similarly treated.

If a farmer collects a sales tax from consumers on sales to them, the tax must be eliminated from his gross sales, either as a deduction or by reporting only net sales as income.

Social Security Taxes

The employer's share is deductible as a business expense. The employee's share is deducted by the employer as part of the wages paid. An employee or self-employed person may not deduct his own Social Security taxes. Unemployment insurance under the new unemployment insurance law will also be deductible as a business expense.

Income Taxes

Federal income taxes paid are not deductible and refunds of federal income taxes are not income. State income taxes and interest on state and federal income taxes on business income and litigating expenses in connection with such taxes are business expenses.

State income taxes attributable to exempt income, other than exempt interest, are not deductible. The portion of the taxes attributable to exempt interest income and to income subject to federal income tax is deductible. Even though gain on a tax-free exchange (§ 3.27) is not recognized for federal purposes, state income taxes assessed upon the gain are deductible.

Special Assessments

Assessments paid for local benefits, such as streets, sidewalks and other improvements are not deductible as taxes nor ordinarily as business expenses. However, the portion of an assessment made for a business purpose of maintenance or repair or meeting interest charges is deductible as a repair or interest.

Excise and Transfer Taxes

Excise taxes, tariff duties, licenses, stamp taxes, and other miscellaneous taxes may be deductible as business expenses or as expenses in connection with property held for the production of income. In farm transactions these are usually part of the cost of goods purchased in the ordinary course of business.

State and local transfer taxes on the conveyance of real estate, securities, and other property are deductible from gross income if the transfer tax is attributable to the taxpayer's trade or business and as an itemized deduction if the transfer tax is paid by an investor in a transaction entered into for profit. However, if the transfer tax is paid in connection with the purchase of a personal residence, it is treated as an addition to basis and if it is paid by the seller on the transfer of a personal residence, it is treated as an expense of sale.

Fees Paid for Tax Advice

Fees and other expenses (e.g., the cost of this book) incurred in connection with the determination, collection, or refund of any tax are deductible. If the matter has a business aspect, it would seem proper to deduct it as a business expense, otherwise it should be an itemized deduction. The IRS would probably say that these are deductible only as itemized deductions.

Farm Income Tax Manual, § 414.

§ 4.10 AUTOMOBILE EXPENSES

Expenses of one or more cars or part of the cars used for business are deductible. If only one car is used solely for business purposes, a separate account is hardly necessary, but if cars are used for business and personal purposes, careful segregation is important.

The deduction may be computed on the basis of fifteen cents per mile for the first 15,000 miles and ten cents per mile thereafter.

To determine the amount of car expenses properly deductible on the business schedule, the total of the automobile expenses for the year is multiplied by the percentage of business use of the car. The percentage of business use is probably best determined by comparative mileage. Hauling produce to town, taking machine parts in for repair, taking home feed or repair parts are business (deductible) uses of the car. Getting the family groceries, going to church, visiting friends, and attending funerals are personal (nondeductible) uses.

Unless a farmer has a truck or another car which is used exclusively for business, it is seldom that his business use of his car is less than 50% of its total use. A great many farm taxpayers have been claiming 60% to 90% use for business, but such claims are being subjected to greater scrutiny. Records of use should be kept.

Personal Use

Where the farmer has two cars and one is claimed as the farm car, the question arises on the extent of the business use of the other car. Any figure should be justified with proof of business use. The best proof is a careful mileage record.

The nonbusiness portions of interest paid and casualty losses suffered on the car are deductible on the personal part of the return. Auto licenses and registration fees are not deductible as taxes unless they are based on value and are, in effect, personal property taxes. See § 4.9.

Commuting costs or nonbusiness travel expenses cannot be treated as farm business expenses. For example, the expense of traveling from home to work is personal and nondeductible.

Items which are solely chargeable to business expense, such as towing or parking charges on a business trip, should be segregated and added separately to the deductible portion of the car expenses.

Car repair costs must be reduced by any insurance received to compensate for damage by collision, fire, storm or other casualty, unless the insurance received was treated as income.

Depreciation should be claimed on the business portion. Where the mileage allowance is used as the measure of the deduction for automobile use, it includes depreciation.

Interest on automobile loans is a part of car upkeep; the principles discussed in § 4.5 apply.

Rental Payments

No deduction will be allowed for so-called lease expenses which in fact constitute payments on the purchase price of autos or trucks. If the lease payments are so arranged as to constitute advance rentals, such payments must be apportioned over the entire rental period.

Auto or truck lease payments are deductible only if they represent ordinary and necessary expenses of, and are directly attributable to, the operation of the farm business. Under no circumstances are such lease payments deductible to the extent that they pay for personal use of a car or truck, for example, vacation trips or commuting from home to farm.

Fifteen Cents Per Mile Allowed

A farmer may compute his transportation costs at fifteen cents per mile for the first 15,000 miles of business travel, and ten cents per mile for business travel in excess of 15,000 miles per year, or he may compute the actual expenses. The mileage allowance is a substitute for all operating and fixed costs of the automobile including depreciation. Parking fees and tolls are separate items. Gasoline taxes for nonbusiness use are separately deducted as a personal expense.

The mileage rate can be used not only on automobiles, but also for panel and pick-up trucks. However, it cannot be used for nonbusiness income-producing activities such as looking after investments.

The mileage rate assumes the use of straight-line depreciation and for purposes of adjustment of the basis of the vehicle, it will be presumed that depreciation was taken on a straight-line basis. The mileage method cannot be used if depreciation has been claimed in the past by any method other than straight-line or if additional first-year depreciation has been claimed.

Farm Income Tax Manual, § 418.

§ 4.11 TRAVEL AND ENTERTAINMENT EXPENSES

In addition to his ordinary costs of transportation in the daily farm business, a farmer, like any other taxpayer is entitled to deduct the costs of business travel (including meals and lodging). Business entertainment may be a proper deduction for some farmers. These deductions may be subjected to a close scrutiny by the Internal Revenue Service, so records and proof are vital.

Both travel and entertainment can be legitimate expenses in any farm operation although entertainment expenses look strange on most farm returns. The general abuse of these deductions, or at least the feeling of the Internal Revenue Service that taxpayers have been abusing them, has led to general suspicion of such deductions. From a practical standpoint, it means that each farmer must be very careful to preserve precise records of travel and entertainment and be prepared to show that the expense is closely related to the farm business.

Business Travel

Travel expenses, including the costs of meals, lodging, tips, laundry, cleaning, and commuting incurred while in a travel status, are properly deductible, so long as they are incurred for business purposes. Cost of operating a private car is deductible, if it is used, although most farmers will have included this under automobile expense.

The cost of any business transportation (not commuting from home to job) is deductible. However, to deduct the cost of meals, the taxpayer must be away overnight from his "tax home." "Home" is regarded as the general area where the business or place of employment is located. Transportation (not meals) is deductible when one works during the day at a temporary or minor location away from the "tax home," for example, a trip to the sale barn. But an overnight trip to buy cattle justifies the deduction of meals, lodging, and transportation.

If a farmer has regular work or business in two or more areas, one must be his tax home. This depends upon the time spent on each job, the business activity in each area and the relative financial return from each. The maximum combination of time, activity and return determines the "home" regardless of where the taxpayer actually lives. His meals and lodging at the place of lesser import are deductible.

A similar rule applies to the farmer who has a separate job, or runs a feed mill separate from his farm, or works two pieces of land in the same general area. He may deduct the cost of transportation between two places of employment both of which produce taxable income.

Because the farm is so often the home, and because the residential farm has become popular, it is not uncommon for the Service to challenge a farmer's transportation to his other place of business as personal, nondeductible commuting.

Clinics, Courses, and Conventions

Travel expenses to farm clinics, short courses and other learning activities that keep one up-to-date are deductible. The theory is that courses taken to maintain one's efficiency and to keep up-to-date are deductible business expenses, while expenses incurred to improve one's self, to seek advancement or promotion or advanced learning are personal, nondeductible expenses.

Convention expenses are deductible if there is a relationship to the farmer's business in the sense that he benefits or advances his interests by attending. Being a delegate is not sufficient justification for deducting expenses. They must be incurred in carrying on the farmer's own business activities. If a spouse's expenses are to be deducted, there must be a business reason for her presence. A personal charitable deduction for out of pocket expenses may be proper if the farmer is a delegate to a church or charitable meeting.

Costs of sight-seeing, entertaining and visiting, unrelated to business, while attending a business convention are nondeductible. Post-convention cruises and tours are personal nondeductible expenses.

If spouse or family attends, the full single cost for the farmer (transportation, lodging, etc.) is deductible and the excess in the double rates, family plan or other package is chargeable to the spouse or family and nondeductible. There is no requirement that a trip be solely for business purposes or even mainly so; rather, it must be an ordinary and necessary expense in the pursuit of, or directly related to, a trade or business.

The 1976 Act added a number of restrictions on deductions related to conventions held outside the United States. While these rules are rather detailed, generally speaking, most expenses related to not more than two such conventions per year may be claimed if all the procedures are followed.

Statement of Travel Expense

A statement should be attached to the return showing (1) the nature of the business in which engaged; (2) the number of days away from home during the taxable year on account of business; (3) the total amount of expenses incident to meals and lodging; and (4) the total amount of other expenses incident to travel and claimed as a deduction.

Details of the expenses are to be kept available for substantiation upon request by the Service. Such requests are often avoided by submitting a statement sufficiently detailed to show that the taxpayer probably does have good records. Receipts should always be saved. In addition, a current memorandum of expenses made each day should show all other expenses. Current cases agree with the Service in denying deductions if the taxpayer cannot show receipts for lodging and for any other expense over $25.

Disallowance of Expenses

No deduction is allowed for entertainment, amusement or recreation unless the expense is "directly related" to the active conduct of the trade or business.

No deduction is allowed for any item with respect to a facility, for example, hunting lodges, fishing camps, swimming pools, cars, planes, hotel suites, and vacation homes, unless it is used "primarily" for the furtherance of the taxpayer's trade or business (generally over half its use) and is "directly related" to the trade or business.

No deduction is allowed for gifts made directly or indirectly to any individual in excess of $25 during the taxable year.

No deduction is allowable for the portion of the living expenses allocable to personal purposes such as sight-seeing or visiting while on a business trip.

If business and personal activities are mixed in foreign travel, part of the transportation costs may be disallowed; for example, attendance at the Turkey Raisers Convention in Edinburgh for three days and a side trip to Rome, Paris and London for twelve will result in disallowance of part of the travel expense from Nebraska to Scotland. Beyond Edinburgh, all expenses are personal. Deduction of the cost of all foreign travel was further limited by the 1976 Act. In no event may the costs of more than two such trips be claimed in a year.

An account book, diary or other contemporaneous record is a must. Receipts are required for items of lodging and for any expense of $25 or more.

Farm Income Tax Manual, § 424.

§ 4.12 DEVELOPMENT EXPENSES

There are two stages in the life of a farm: (1) pre-productive and (2) productive. During the first period, the farmer has the option to deduct or to capitalize ordinary and necessary expenses. During the productive period, these expenses must be deducted. At an earlier time, the Service argued that there were preparatory, development, and productive stages. The first two of these have now been collapsed into the pre-productive. This change in terminology does not represent any change in the deductibility of expenditures.

The costs of citrus and almond groves for the first four years must be capitalized.

For rules for farm syndicates and certain corporations, see § 4.1.

The Service now divides the life of a farm into two periods:

(1) the pre-productive and

(2) the productive

Previously, the Service argued there was a **preparatory** period and a **development** period, but they are now treated as the single pre-productive period.

Pre-Productive Period

The change in terminology was not intended to change the law about deduction of expenditures. Costs, including labor and materials, for capital items are not deductible; for example: the cost of clearing brush, trees and stumps; leveling and conditioning land; trees and tree planting; drilling and equipping wells; building irrigation canals or ditches; laying pipes; installing drain tile; straightening creek beds;

constructing tanks, reservoirs or dams; building roads; and similar items.

The costs of items which are depreciable will be recovered over their useful lives but other capitalized expenses will be recovered only through depletion (in the case of timber) or only on the sale of the land because the expense is added to basis and land is not a depreciable asset.

There is an election to capitalize interest, taxes and carrying charges during this period. These are items which would normally be deductible.

The special elections to deduct soil and water conservation expenses, § 4.13, and the cost of fertilizer and lime, § 4.7 are not available until the land has been "used in farming." Land never previously used for farming does not qualify. Soil and water conservation expenses in excess of 25% of gross income from farming for the year may not be deducted, and there may be no gross income from farming unless there is another source of farm income. Excess conservation expenses may be carried over to future years, but the others may not be. However, expenses for clearing land to make it suitable for farming may be deducted but the limitation of the lower of $5000 or 25% of taxable income from farming may make the election worthless. See § 4.14.

See §§ 5.12 and 5.13 on recapture as ordinary income of soil and water conservation expenditures and land-clearing expenses if the land is sold within ten years of acquisition.

After planting or similar activity, a farmer has an option to capitalize ordinary and necessary business expenses. He must care for his orchard, which does not yet produce income. He must cultivate and spray the trees, pay for the water he uses for irrigation, and buy fertilizer which has to be replaced each year. These "cultural practices expenditures," may be deducted or capitalized at the taxpayer's option. However, once the election to capitalize is made, the taxpayer may not change it on an amended return.

As noted below, these rules do not apply to citrus (1970 and thereafter) or almond (1971 and thereafter) groves, nor after 1975 to any orchard or vineyard held by a farm syndicate. Further, under the 1976 Act, some corporations are required to use the accrual method of accounting and must capitalize pre-productive expenses. See § 4.1.

The fact that development expenses may be deducted or capitalized has nothing to do with conservation or land-clearing expense elections because they cover expenses which could not be currently deducted except by making the specific elections.

Productive Period

When the farm, orchard or ranch reaches its productive period, the farmer loses any option he had during the pre-productive period. The farm is now a full-fledged operating business. Ordinary and necessary expenses cannot be capitalized.

Instead the farmer must deduct these expenses each year, either when they are incurred, if he uses the accrual method, or when they are paid, if he uses the cash method.

Of course, during the productive period, the conservation, land-clearing and fertilizer expense elections are applicable.

Special Rule for Citrus and Almond Groves

Beginning in 1970 for citrus and in 1971 for almonds, costs of groves must be capitalized for four years. Any amount which would have been allowable as a deduction under the rules above and which is attributable to the planting, cultivation, maintenance or development of the grove must be added to capital account if it is incurred in the taxable year in which the trees are planted or in the next three taxable years. If part of a grove is planted in one year and part in another, the four-year period runs on each portion from the year the trees were planted. This rule does not apply to a grove which was replanted after having been lost or damaged by freeze, disease, drought, pests or casualty or which was planted or replanted before December 30, 1969 (for citrus) or before December 30, 1970 (for almonds).

See § 4.1 for new rule on farm syndicates and farm corporations.

Farm Income Tax Manual, § 419.

§ 4.13 SOIL AND WATER CONSERVATION EXPENSES

Soil and water conservation expenses may, if the farmer elects, be treated as currently deductible expenses up to 25% of gross income from farming in any year. Expenses exceeding 25% of gross income may be carried forward to future years within the limitation of 25% of gross income from farming in each such year.

Deductions for soil and water conservation expenditures may be recaptured as ordinary income on the sale of farm land within ten years of acquisition.

A taxpayer, engaged in the business of farming may deduct the expenditures for soil and water conservation in respect of land used in farming or for the prevention of erosion of land used in farming. Unless he so elects, these expenses are not deductible nor even recoverable through depreciation. A $1000 expense for bulldozing is simply an addition to the cost or basis of the land which will save nothing in income taxes until the farmer sells the land.

The election is made without consent in the first year in which the farmer has any soil and water conservation expenditures. The treatment of these expenses (i.e., deduction or failure to deduct) in the first year they are incurred amounts to an election which will be binding on the taxpayer for future years. If he paid or incurred any such expenditures before the current taxable year, he has, knowingly or unknowingly, made his election. Once a method is adopted it must be adhered to for that year and subsequent years unless the District Director authorizes a change as to part or all of such expenditures.

The election to deduct the expenditures may be made with respect to a single project or single farm and may be limited to that. This requires consent of the District Director whether it is the first expenditure or whether it represents a change in a previous election.

Deductible Expenses

The scope of soil and water conservation activities is broad. It includes any detention or water-control activities, filling gullies, collecting water, retarding runoff and prevention of flood damage. Among the expenditures deductible are moving of earth; leveling, grading and terracing; contour furrowing; the construction, control and protection of diversion channels, drainage ditches, earthern dams, watercourses, outlets and ponds; eradication of brush; and planting of windbreaks. A recent ruling holds that the costs of irrigation ditches to bring water from outside the farm area are soil and water conservation expenditures.

Also included are assessments levied by a soil or water conservation or drainage district to the extent these expenditures would constitute deductible expenditures if made directly by the farmer. A limited form of amortization over ten years is allowed for assessments to purchase depreciable property.

Items, or parts of items, that are depreciable must be depreciated and may not be charged off as a soil and water conservation expense. Any expense otherwise deductible is not a soil and water conservation expenditure.

Thus, excluded are: (1) the purchase, construction, installation or improvement of structures, appliances or facilities which are of a character subject to depreciation, or (2) amounts paid or incurred which would be allowable as a deduction anyway.

The tiling of land may be for the purpose of soil conservation or to prevent erosion but the expense would be depreciable and therefore would not be deductible as a conservation expense. Ponds dug without the use of depreciable materials would be fully deductible although the pipes and pumps which might be used in connection with the pond would be depreciable and not currently deductible.

Expenditures for taxes, interest or repairs related to soil and water conservation facilities are otherwise deductible. So are costs of seed, fertilizer, lime, planting stock, and similar expenses, but if these are used in a gully stabilization project, they would be soil and water conservation expenses because they would not otherwise be deductible.

The part of the purchase price of a farm allocated to earthern watering tanks of indeterminate life constructed by the prior owners of the land was not recoverable by the purchaser either as a soil and water conservation expense or through depreciation.

Business of Farming

The "business of farming" is engaged in by an owner or tenant who cultivates, operates or manages a farm and a landlord who receives a share rent. However, a landlord who receives a "fixed rental," not based upon a share of production, is not in the "business of farming" and cannot elect to deduct these expenses unless he "materially participates" in the operation or management of the farm. See § 7.5.

Forestry or the growing of timber is not the business of farming. Nor is the hobby or recreational farmer in business. "Farm" is used in its usual sense to include stock, dairy, poultry, fruit and truck farms and plantations, ranches, ranges and orchards and fish farms. A person is in the business of farming if he is a member of a farm partnership engaged in the business of farming.

"Land Used in Farming"

"Land used in farming" includes land used by the farmer or his tenant for production of crops, fruits or other agricultural products or for the sustenance of livestock, including fowl.

Land must be presently used (or have been used some time in the past) for farming to qualify for the election. Expenditures on newly acquired idle land are not deductible. However, land is "used in farming" if it is part of a farm even though temporarily idle as a conservation practice.

Newly acquired land is considered as used in farming if the previous owner used the land for farming and the new use is substantially a continuation of the old. If the taxpayer is preparing the land for a different use or for its first farming use, the land is not land used in farming for purposes of deducting soil and water conservation expenses. Such costs should be capitalized and added to the basis of the farm. See § 4.12.

"Gross Income from Farming"

The amount deductible cannot exceed 25% of the gross income from farming in the taxable year. To the extent the expenditure exceeds

25% it may be carried forward to succeeding taxable years, but the amount of the deduction for the carry-over and the new expenditures in that succeeding year may not exceed 25% of gross income from farming in that year. The carry-over is considered the first expenditure for the following year. In a partnership, the 25% limit applies to each partner, not to the partnership.

"Gross income from farming" is the gross income derived from the business of farming as defined above. It includes income from farms on which no expenditures for soil and water conservation were made. Under the definition, it would not include the income from land which produced cash or fixed rents unless there was material participation. Nor does it include gains or losses from the sale of Section 1231 items or other assets used in the business except livestock.

Gains from sales of draft, breeding, dairy, and sporting livestock are included in gross income from farming while gains from other assets used in the business are excluded. The livestock gains are not reduced by the 50% deduction for long-term gains.

Carry-over

Excess of deductions over 25% of gross income from farming in any one year may be carried over to subsequent years indefinitely. The carry-over is deducted first against the 25% allowed limitation for the succeeding year and any accumulated excess is carried forward to the next succeeding year. The trap lies in the fact that the carry-over deduction is measured by farm income and the life of the taxpayer. Thus, if a farmer retires or dies with a carry-over, it is lost. The retired farmer will no longer have gross income from farming, and an estate of a decedent cannot take the decedent's deduction. If a corporation terminates, the carry-over is lost. Termination of a partnership does not lose the carry-over because it belongs to the partner, not the partnership.

No adjustment can be made to the basis of land for the amount of the unused carry-over. Thus, the lost deduction is not recovered on sale of the farm. If the taxpayer ever gets back to farming, the carry-over is still there to be deducted.

Part of Net Operating Loss

The soil and water conservation deduction can be a part of a net operating loss which can be carried back or forward. Once the soil and water expense becomes part of the net operating loss, it loses its identity and is not subject to the 25% limitation in the year to which the loss is carried. Since the soil and water conservation expense deduction for any one year is limited to 25% of gross income from farming, only this much could go into the computation of the net operating loss. The excess is carried forward to the next year as a soil and water expense carry-over.

Recapture of Deductions

Soil and water conservation expenditures may be recaptured as ordinary income on the sale of farm land at a gain if the land is sold within ten years of acquisition.

If a farmer has an "excess deductions account," on the sale of farm land the deductions for soil and water conservation expenditures (and land-clearing expenses) claimed in the taxable year and the prior four years will be recaptured as ordinary income to the extent of the gain on the land and the balance in the EDA.

Even if there is no EDA, there will be recaptured as ordinary income some or all of the soil and water conservation expenditure (and the land-clearing expense) deductions if the land is sold within ten years of acquisition. If the land is disposed of within five years of acquisition, all of the deductions are recaptured. If the land is disposed of within the sixth year, 80% of the deductions are recaptured; within the seventh year, 60%; within the eighth year, 40%; and within the ninth year, 20%. The amount to be treated as ordinary income is the lesser of the gain on the sale or the applicable percentage of the deductions claimed. One point to note is that the percentage of recapture is based on the period that the land has been held. Thus, if the land has been owned eleven years prior to sale, there is no recapture of these expenses even though they may have been made much more recently, for example, in the last year or two before sale.

There is an overlap between the two provisions. The deductions are first recaptured against the EDA, if there is one, and any excess is recaptured under the ten-year provision. See §§ 5.12, 5.13.

Farm Income Tax Manual, § 420.

§ 4.14 LAND-CLEARING EXPENSES

Expenditures for land clearing for the purpose of making land suitable for use in farming may, if the farmer elects, be treated as currently deductible expenses up to $5000 or 25% of taxable income derived from farming in any one year. There is no carry-over of excess expenses to future years.

Deductions for land-clearing expenses may be recaptured as ordinary income on the sale of farm land within ten years of acquisition.

Expenditures for clearing land may be deducted if they are for the purpose of making the land suitable for use in farming. The deduction is limited in any taxable year to the lesser of $5000 or 25% of the taxable (not gross) income derived from farming during the taxable year. Expenditures in excess of the allowable limits must be capitalized and added to the basis of the land. There is no carry-over to subsequent years.

Land-clearing expenses do not include expenditures for structures or facilities which are depreciable, nor expenditures that would properly be soil and water conservation expenses nor any expenditure that would otherwise be deductible. Depreciation on equipment used to clear land is deductible only as a land-clearing expense, not as depreciation.

Definitions

"Clearing of land" includes (but is not limited to) removal of rocks; eradication of trees, stumps and brush; the treatment or moving of earth; the diversion of streams and watercourses; and the removal of salt or other minerals.

"Taxable income" is gross income from farming reduced by all proper farm deductions, including deductions for soil and water conservation and fertilizer expenditures which the farmer elected to deduct but excluding the land-clearing expenses. It does not include gains or losses on business assets (except livestock) or land. Gains and losses from the sales of draft, breeding, dairy and sporting livestock are included; gains are reduced by the 50% capital gains deduction. Taxable income means all farm income of the taxpayer farmer including income from farms upon which no clearing is done.

The "business of farming" assumes an owner or tenant operating the farm for profit; a landlord receiving a rental not based upon a share of production is not in the business of farming unless he materially participates in the farm operation. Forestry or the growing of timber is not the business of farming.

"Land suitable for use in farming" means land which, as a result of the clearing, is suitable for use by the farmer or his tenant for the production of crops, fruits, and other agricultural products or for the sustenance of livestock. The deduction does not apply to expenses to clear land for sale or for a nonfarming use.

How to Make Election

The election is made each year by a signed statement attached to the return showing the expenditures and the computation of the limitation. If the election is not made, the expenditures are capital and are added to the basis of the land. But an election for one year does not bind for subsequent years.

A partnership makes the election to deduct expenditures for clearing land, but each partner takes into account separately his distributive share of the partnership's expenditures for clearing land. If the partner has land of his own upon which he has elected to deduct land-clearing expenditures, he must aggregate his own "taxable income derived from farming" with his share of the partnership's similar income and aggregate his separate expenditures and his share of the partnership expenditures for clearing land.

Recapture of Deductions

Land-clearing expenses may be recaptured as ordinary income on the sale of farm land at a gain if the land is sold within ten years of acquisition.

Upon sale of farmland, a farmer who has an "excess deductions account" will recapture deductions for land-clearing expenses (and soil and water conservation expenditures) claimed in the taxable year and the prior four years as ordinary income to the extent of the gain on the land and the balance in the EDA. No additions are made to the EDA after 1975.

Under another provision, which applies even if there is no EDA, there will be recaptured as ordinary income some or all of the land-clearing expense (and soil and water conservation expenditure) deductions if the land is sold within ten years of acquisition. If the land is disposed of within five years of acquisition, all of the deductions are recaptured. If the land is disposed of within the sixth year, 80% of the deductions are recaptured; within the seventh year, 60%; within the eighth year, 40%; and within the ninth year, 20%. The amount to be treated as ordinary income is the lesser of the gain on the sale or the applicable percentage of the deductions claimed. One point to note is that the percentage of recapture is based on the period that the land has been held. Thus, if the land has been owned ten years prior to sale, there is no recapture of these expenses even though they may have been made much more recently, for example, in the last year or two before sale.

There is an overlap between the two provisions. The deductions are first recaptured against the EDA, if there is one, and any excess is recaptured under the ten-year provision. See §§ 5.12, 5.13.

Farm Income Tax Manual, § 421.

§ 4.15 ADJUSTMENT OF EXPENSES FOR PRODUCE CONSUMED

The value of the farmer's own produce consumed by himself and family should not be reported as income but the expenses of producing the consumed produce are not deductible.

Ordinarily, the matter of adjusting the farmer's expenses to exclude costs incurred for personal, nonbusiness purposes can be taken care of to a considerable extent by simply omitting items from expense accounts. For example, by excluding garden seeds from seed costs practically all the expense connected with putting vegetables on the farmer's table are thereby removed from his cost accounts. By leaving out suitable amounts from feed and other costs, his expenses claimed on the return will reflect only those expenses incurred for business reasons. In so far as this is not otherwise accomplished, the farmer should adjust his total of listed expenses by the cost to him of producing what he and his family consumed.

It is impossible to say whether this adjustment should amount to $50 per year per person, $100 per year, or $200 per year. The problem is similar to determining the amount it costs a farmer to feed a hired man. There is a wide variation among individuals in the amount of home-produced food they use, the amount bought in grocery stores, and the costs to them of producing food. A calf butchered by the farmer might involve almost negligible cost if raised on the farmer's own pasture or fed with hay and corn raised by the farmer, but might involve a considerable expense if raised on high-priced commercial feeds.

The Practical Answer

If the District Director adopts a local standard to reduce farm business expenses for the costs of produce consumed, the farmer must accept that standard or show by his own records that the Director's figure is too high. The farmer, therefore, should fortify himself with such evidence as will bear on the matter. The state agricultural university will have some helpful figures on production costs.

The pragmatic answer is the same as that suggested with respect to salvage value. See § 5.7. If the farmer makes no entry for consumption of farm products, it is extremely difficult to challenge an auditing agent's figures because the farmer usually has no substantial proof of the costs of raising. However, if the return shows on its face some reasonable figure by which the deductions have been reduced to account for home-consumed produce, the chances are very good that no issue will be raised.

Farm Income Tax Manual, § 425.

§ 4.16 LOSSES GENERALLY

Most losses incurred in the operation of farms as business enterprises and losses incurred in any transaction entered into for profit are deductible from gross income. Special types of losses are discussed separately:

Losses incurred in normal farm operations or in profit transactions will appear as an excess of deductions over income. Otherwise, losses result from and are realized by the happening of an event that closes the loss transaction and makes the amount of the loss determinable.

Farmers usually know when a loss occurs because most losses result from events such as a sale, storm, or fire. However, in the case of a joint venture, no loss is realized by the joint venturer until salvage operations are completed.

Losses of the farmer on sales and exchanges of ordinary stock in trade will require no particular spot on the tax return as they will show up automatically as sales and expenses are reported in their proper places. When business assets and other Section 1231 items are involved they will be reported as indicated in § 3.12. Capital losses will be reported on Schedule D, filed with Form 1040. See § 3.12.

See § 4.21 for nondeductible losses.

See § 4.22 for disallowance of losses on hobby farms.

No Loss of Anticipated Income

Deduction of losses of anticipated income is a strangely recurring problem. Unless an item of anticipated income has been reported on an income tax return (this would normally happen only in the case of an accrual-basis taxpayer) there is no loss of any nature for the failure to obtain the income. For example, the loss of a raised cow or a growing crop by a cash-basis farmer who deducted all of the costs of raising produces no deductible tax loss, even though there is a very real economic loss. A loss for tax purposes may never exceed basis.

Farm Income Tax Manual, §§ 426, 427.

§ 4.17 CASUALTY AND THEFT LOSSES

Casualty and theft losses are deductible whether the property is used in the trade or business, held for the production of income, or held for personal use. The losses may be fully deductible losses or Section 1231 losses. See § 3.12. Casualty and theft losses may

become a part of a net operating loss carry-back and carry-forward. See § 4.18.

A casualty loss is the complete or partial destruction of property resulting from an identifiable event of a sudden, unexpected or unusual nature. Theft is the criminal taking of money or property or embezzlement. Casualty and theft losses are deductible whether the property is held for business purposes, for investment, or for personal purposes. However, the measure of the loss and the manner of reporting may vary depending upon the use of the property.

Any casualty loss of business property is a business loss as such, but a loss that is not treated as a casualty loss may still be a deductible business loss—although the time of deduction may be delayed. The Internal Revenue Service uses the same interpretations to determine when a business loss occurs as a result of casualty as it does in defining the deduction for casualty losses from property not used in the trade or business.

For example, the drying up of a well is not a casualty loss because it does not result from sudden destructive force nor is it an unusual or unexpected event, but the loss may be deductible as a retirement or abandonment loss.

Termite damage is not deductible as a casualty loss because it is not sudden, but losses of business property resulting from termite damage are deductible business losses. Too, on business property and property held for the production of income, a loss that would not qualify as a casualty loss will ultimately be picked up on sale or absorbed through depreciation.

Lost Property

Business property which is lost or strays away may result in a deductible business expense, even though not a casualty or theft loss. Loss of business property such as machine parts which drop off while the machine is being moved, or a purchased hog which strays away and is never found, is deductible by the farmer as an ordinary business loss. This is true even though the farmer is unable to prove that the loss was a "casualty" or can furnish no proof that the item was stolen.

This is not true of losses claimed as personal deductions, for example, loss of a watch or ring or a valuable house dog. In such cases the taxpayer cannot successfully claim a loss unless he can prove that the property was stolen or destroyed by casualty.

The unauthorized pledging of collateral held as security for a loan is a theft loss. Legal fees paid to recover moneys converted by another are deductible as a theft loss.

Nature of Loss

The definition of casualty prevents a current loss deduction because of deterioration or nonphysical influence. Progressive deterioration such as the weakening of foundation walls through normal wind and weather conditions is not a casualty. Loss of value because property is close to a disaster area and there is a possibility of recurrence of the disaster is not a casualty loss.

Loss of trees from disease or insects is not sudden enough to be a casualty loss. Thus a casualty loss is allowed when the damage is of a physical nature resulting from accident, fire, storm, shipwreck, flood or other sudden destructive force.

Sudden and severe frost may result in a casualty loss but the Government claims that prolonged drought cannot produce a casualty loss. An earthslide caused a casualty loss even though the contractor had been warned and negligently failed to protect against the possibility of the land shifting.

Loss of useful value without physical damage is not deductible as a casualty loss even though it results from a casualty and the probability of its recurrence. Physical damage or destruction must be shown to prove a casualty loss. Future losses or depreciation of property are not currently deductible.

Since the deductibility of a loss turns upon an identifiable event which results in the realization of the loss, a decline in value of property would not be a business loss until the property is sold, unless physical damage resulted from a casualty. For example, the decline in value resulting from buyer resistance because property is in an area subject to storms, landslides, or other casualty is not a deductible loss.

When Deductible

Ordinarily, casualty losses are deductible in the year in which the casualty occurs. If the amount of the loss is not ascertained by the end of the year, the taxpayer should request an extension of time to file the return or estimate the loss on his return. If the estimate is incorrect, he should file an amended return for the year of loss.

A loss is regarded as incurred in the year in which it can be measured. A delayed reaction to a storm or a hurricane or other casualty is deductible when the actual damage appears. Theft losses are deductible only in the year that the theft or embezzlement is discovered.

Special Rule for "Disaster" Areas

Casualty losses suffered by property owners in so-called disaster areas (which are so declared by the President) may be deducted on the tax return for the year prior to the casualty loss. For example, a loss from a disaster casualty in the area designated by the President suffered at any time in 1977 may be deducted on the tax return filed for 1976, either as originally filed or as subsequently amended. In effect, this gives a quick tax benefit to the sufferer.

Lists of disasters and disaster areas are reported in the Internal Revenue Service Bulletin which is issued weekly.

This is not an additional deduction. It is an election to treat a casualty loss suffered in one year as if it had taken place in the prior year which has closed.

Measure of Loss

Most casualty loss problems arise in determining the amount of the deductible loss. Failure to prove the loss and failure to account for compensation are the common hurdles.

The measure of a total loss on business property or property held for the production of income is the adjusted basis of the property, regardless of value.

The deduction for a partial loss of the property is the difference between the fair market value before the casualty and the fair market value after the casualty, but in no event may the deductible loss exceed the adjusted basis of the property. In effect, the measure of the loss is the actual enconomic loss suffered but not in excess of the basis of the property.

The difference in the value of the property immediately before and immediately after the casualty should be supported by an appraisal made by an experienced and reliable appraiser. The appraisal fees are deductible.

Repairs and cleaning-up costs after a casualty are ordinary business expenses. If this is the measure of the loss and the property is thus restored, the deduction may not be claimed twice. If the repairs actually improve the property beyond its condition prior to the casualty, the costs of the improvement are not part of the casualty loss.

Expenditures for protection against future casualties are not deductible but are treated as new capital expenditures. For example, a dike or levee to prevent further flooding is not part of the deductible flood loss.

Insurance Proceeds

The casualty loss must be reduced by insurance or other compensation received. The Service has always maintained the position that the amount of the loss is figured first (the adjusted basis or the loss in value, whichever is lower) and then the insurance or compensation is applied to reduce the deduction.

For example, if a $10,000 barn with a basis of $5000 was burned down (an economic loss of $10,000 but a tax loss of $5000) and insurance of $5000 was collected, there would be no deductible loss.

If the insurance has not been received at the end of the year, the loss should be reduced by the estimated recovery. An amended return is filed when the final figures are known.

Rehabilitation payments received for restoring or rehabilitating property lost or damaged reduce the amount of the deductible loss. If the reimbursement exceeds the basis of the property, the amount of the excess cannot be used to increase the basis of the property. However, gifts made to hurricane victims to tide them over difficult times did not reduce the casualty loss even though the money was used to rehabilitate the property because the gifts did not have to be so used.

Growing Crops or Raised Livestock

Cash-basis farmers have no deductible loss on growing crops if all of the costs of raising have been deducted. Losses of inventory property are reflected in the inventory and no separate loss is claimed unless the opening inventory or purchases are adjusted. Cash-basis taxpayers have no casualty loss on raised animals if they deducted all of the costs of raising. On purchased animals, the loss is limited to the adjusted basis of the animal.

The measure of the loss of shrubs and ornamental trees on business property is the difference between the value of the trees (not the property), before and after the casualty, but not more than the basis of the trees or shrubs.

Deduction has been allowed for loss of topsoil from floods, for topsoil blown away by a storm, for the loss of purchased sod and asparagus ferns by frost, and for sand washed away from a man-made beach.

Termites do not work fast enough to create a casualty loss but the cost of efforts to counteract termites, repair of damage and the loss when the structure falls down are business expenses. The decline in value caused by the termites is not deductible.

Limitation on Losses

The amount allowable as a deduction for a casualty loss resulting from destruction of timber by a hurricane may not exceed the adjusted basis of the quantity of timber which by fair and reasonable estimates is found to be unfit for use by reason of the hurricane. Such adjusted basis does not include any portion of the basis attributable to the land, other improvements or to any timber not rendered worthless by the hurricane.

The Service requires that the portion of the timber subject to hurricane damage be segregated from the entire tract or tracts. The measure of the loss is (1) the fair market value of the destroyed timber, or (2) the adjusted basis of the destroyed timber, if lower.

Where there is damage to different kinds of business property, the loss must be computed separately for each single identifiable unit. Thus, a casualty loss to a citrus grove must be determined separately as to the trees and to the land. The measure of a total loss of the trees is the fair market value of the trees prior to the casualty but not in excess of the adjusted basis of the trees. There is no loss deduction for the fruit if it has no basis, and in the usual case, it will not. An accrual-basis taxpayer who followed the 1976 rulings which require growing crops to be inventoried would have a basis for citrus fruit included in an inventory at year's end.

There could be a loss to the land. The measure would be the difference between the value of the land and the value of the land with dead citrus trees.

Full Loss or Section 1231 Loss

Section 1231 includes involuntary conversions which are defined to include losses from destruction or theft of property used in the trade or business or capital assets held for over six months in 1976, nine months in 1977 and one year thereafter, and livestock held for draft, breeding, dairy, and sporting purposes and held for 24 months (cattle and horses) or twelve months (other livestock).

Gains and losses from casualties and thefts are first aggregated to see whether there is a net gain or loss. If there is a net gain from casualties and thefts, all of the transactions are included with other Section 1231 transactions. If there is a net loss on the casualty and theft aggregate, the transactions are not included with Section 1231 transactions. Thus, if there is a net gain, the gain will be capital or it will offset Section 1231 losses, while if there is a net casualty loss it will be an ordinary loss.

Assets which are held for a lesser period and livestock held for draft, breeding, dairy, and sporting purposes for less than 24 months (cattle and horses) or twelve months (other livestock) produce ordinary casualty losses not included either in the casualty and theft aggregate or under Section 1231.

The basis of property damaged by casualty is reduced by the amount of the deductible loss. The basis is increased by any amounts spent to rebuild the property.

If the insurance or other recovery exceeds the adjusted basis of the property damaged or destroyed by a casualty, the taxpayer has realized a gain. The gain is ordinary or capital according to the nature of the property, but the tax on the gain may be postponed if the proceeds are invested in replacement property. See § 3.24.

Net Operating Loss

A casualty or theft loss of property held for business, investment, or even personal purposes, including, for example, a residence or a pleasure boat, becomes a part of the net operating loss which may be carried back and forward. See § 4.18.

No Floor on Business Property

There is a limit on the amount of the deduction for casualty and theft losses of property held for personal purposes, which does not apply to business property or property held for the production of income. Personal casualty losses are deductible or added to the aggregate of casualty transactions only to the extent that the loss from each casualty or theft exceeds $100.

Farm Income Tax Manual, §§ 428, 429.

§ 4.18 NET OPERATING LOSSES

A net operating loss may be carried back for three years to offset income in those years. Unused portions of the carry-back loss may be carried forward for seven years. Losses on the sale of farm land and farm business property are included in the net operating loss.

Casualty losses on nonbusiness properties are also included in the net operating loss to be carried back and forward.

A taxpayer may elect not to carry the loss back but to carry it forward only.

A net loss from the operation of the farm during the taxable year is fully deductible from the farmer's income from any other sources, provided the farm is not operated for recreation or pleasure. To the extent that there remains a net loss after other gross income for the year has been set off, the loss can be carried back successively to the third, second, and first years preceding the year of the loss. The taxpayer may choose, however, not to carry the loss back but merely use if as a carry-forward. If the loss is carried back, it is offset first against the income of the third preceding year. Any amount of the loss not absorbed by carry-back to the third preceding year is carried to the second preceding year and then to first preceding year. If the loss is not used up by the carry-back, it is carried forward to the years after the year of the loss in chronological order for seven years.

Loss on the sale of the taxpayer's farm is treated as a farm operating loss. Any losses on the sale of business assets (depreciable property and real estate) are business losses and become part of a net operating loss.

A casualty loss on property, whether or not connected with the trade or business, is included in the net operating loss. Thus, even a fire loss on a residence becomes a net operating loss subject to the carry-back and carry-forward.

Computing the Loss

An excess of deductions over gross income is a loss. However, the net operating loss is computed with certain modifications. The modifications (omitting those which are not likely to concern a farmer) are these:

1. Net operating loss deductions from another year cannot be added to the current year's loss.
2. Losses on sales or exchanges of capital assets may not exceed gains, and to the extent capital gains were reduced by 50% in computing income, this 50% must be restored to income.
3. No deduction for personal exemptions is allowed.
4. Nonbusiness deductions are allowed only to the extent of nonbusiness income.

Adjustment for Capital Gains

A farmer, in determining the amount of the net operating loss to carry back, must reduce the loss by the 50% deduction taken into account on capital gains. For example, a farmer may have sold $2000 worth of raised brood sows and $3000 worth of raised dairy cows for a long-term capital gain of $5000. In computing his taxable income he deducted $2500 by applying the capital gain provisions to the sale of his dairy and breeding animals. These gains must be offset in full before any loss can be available to carry back.

When the loss is carried back, no adjustment is made for capital gains in the first year to which the loss is carried. If the loss is in excess of the taxable income of the first year to which it is carried, so that some part of the loss will be carried over to a subsequent year, the capital gain adjustment arises again. The amount of the loss that may be carried over to the next year is reduced by the amount of the capital gain that was not taxed.

For example, if 1976 produces a net operating loss of $5000, it will be carried back first to 1973 unless the taxpayer chooses to carry it forward only and not back. Assume that 1973 taxable income was $6000 made up of $5000 ordinary income and $2000 of long-term capital gains less the deduction for 50% of the capital gains. Application of the 1976 loss to 1973 would result in taxable income for 1973 of

$1000. The loss is consumed and there is no further carry-over. If the 1976 loss were $8000, application of the loss to 1973 would wipe out 1973 income and leave $2000 of loss. In determining the amount of the loss that may be carried to 1974, the $2000 excess loss must be reduced by the $1000 of 1973 capital gains which were not subject to tax. Thus, the carry-over to 1974 is $1000.

Nonfarm Income

If a farmer suffers a loss on the sale or operation of his farm, his income from rents, wages and other business sources is applied to reduce the loss. Business income is not reduced by nonbusiness deductions. For example, if the farm operating loss was $1000, salary earned was $1000, and nonbusiness deductions were $1000, the salary is applied to the operating loss first, so there is no loss to carry back. The unused nonbusiness deductions give no tax benefit. Nonbusiness deductions offset only nonbusiness income.

A wife's salary on a joint return is business income which is applied to reduce the net operating loss. The gain from the sale or other disposition of depreciable or real property used in the trade or business is trade or business income which reduces the operating loss.

The salary or wages paid to a shareholder of a Subchapter S corporation is business income which must first offset the net operating loss, but the shareholder's prorata portion of the undistributed corporate income or amounts distributed to him as dividends is nonbusiness income.

Income Against Which Loss Applied

In the year to which the net operating loss is carried, it is offset against the taxable income of that particular year computed without the deduction for personal exemptions. Further, the taxable income for the year to which the net operating loss is carried must be recomputed if there was any deduction which was based upon a percentage of taxable income or adjusted gross income.

Quick Refunds

Form 1045 for an individual or Form 1139 for a corporation is used to obtain a refund for the year to which a net operating loss is carried back. The Internal Revenue Service will ordinarily act on the application within ninety days from the date on which it was filed although the determination of the refund is subject to later examination.

Farm Income Tax Manual, § 430.

§ 4.19 ABANDONMENT, RETIREMENT, AND DEMOLITION LOSSES

Sudden loss, for unforeseen reasons, of useful value of an asset used in the business results in a deductible loss. Retirement of assets depreciated on the basis of their maximum useful life results in a deductible loss. Demolition of a building may result in a loss or an addition to the basis of the land.

Where the usefulness of an asset used in the business suddenly terminates because of changed conditions so that the farmer discontinues the use of the asset and discards it permanently, a deductible loss results.

An abandonment or retirement loss generally may not be claimed if the loss of useful life terminates solely as a result of the gradual processes for which depreciation is allowed. However, if assets are depreciated at a rate based upon maximum useful life, a deduction for early retirement of the asset is allowed based on the difference between the adjusted basis of the asset and its salvage value.

Most of the troubles in claiming abandonment losses turn on proof of abandonment. Abandonment is total disassociation from the asset. The Service demands convincing proof of the permanency of abandonment. For example, no loss was allowed when, at the time of the "abandonment," the taxpayer intended to sell the property and later did sell it. An abandonment loss is regarded as an exception to the normal rule that the realization of a loss requires a sale or disposition of the property.

Claiming such loss requires proof of the unforeseen circumstance causing the property to be prematurely discarded. Since such a loss is usually difficult to prove, it is frequently wise to sell the asset as junk so that the loss can be accurately determined.

Assets Subject to ADR

If a taxpayer has elected the advantages of the Asset Depreciation Range System, discussed at § 5.6, he may be at a disadvantage in claiming an abandonment loss for assets depreciated under ADR. In those cases a loss may be claimed only if the loss results from an "extraordinary" retirement. A retirement is extraordinary only if it results from (1) a casualty or (2) a curtailment or cessation of a business or an identifiable part of a business and such curtailed assets amount to 20% of the assets in the particular asset account. Obviously, this set of rules is not as liberal as the ones discussed above.

Water Wells

Amounts expended in the drilling of test holes are capital expenditures and are not deductible. If a sufficient supply of water is found and a water well drilled, abandonment of the test holes does not give rise to a deduction. The cost of the test holes is a part of the cost of the well. If the well or any of its components has a determinable useful life, all or part of the cost is depreciable.

If sufficient water is not found after the drilling of the test holes and they are abandoned, a loss is sustained and the entire cost is deductible when the project is abandoned, even though the test holes may have been drilled in different taxable years.

Abandonment of a well for bona fide business reasons indicated by some overt act or identifiable event, such as filling or sealing, which terminates the existence of the well, results in a deductible loss. If the cost of the well or its components was depreciated, the loss is the amount of the undepreciated basis.

Demolition

Demolition of buildings to make way for new structures may or may not result in a deductible loss. Sometimes a farmer acquires land with buildings on it and then demolishes the buildings. If he has no allowable loss, the adjusted basis of the buildings demolished and the costs of demolition are added to the basis of the new structure or the land.

Generally, if the land is acquired with an intent to demolish the buildings, no deduction is allowed for loss on the buildings. The cost of the buildings and the cost of demolition is added to the basis of the land. If the buildings are in fact used, a portion of the basis of the property is allocated to the buildings for depreciation. The allocated portion of the purchase price is the present value of the buildings measured by the rents which could be realized over the period of intended use. The unrecovered portion of the basis so allocated is a loss on demolition or is added to the basis of the land.

If the intent to demolish is formed after acquisition, there is a loss equal to the remaining basis plus the cost of demolition.

There is a pattern in some areas of buying land with buildings to get the land, and the buildings are soon removed. In this case, it seems advisable to allocate a reasonable value to all salable or reusable materials (so that there is a basis for sale or for structures made from the old materials) and treat the rest of the cost plus the demolition expense as basis of the land.

If the building to be demolished is an historic structure, there will be no loss allowed unless a number of technical rules are followed. See § 1.22.

Farm Income Tax Manual, § 431.

§ 4.20 BUSINESS BAD DEBTS

A debt which arose in the farmer's trade or business which becomes worthless in whole or in part may be deducted. But the debt must be bona fide and not a mere failure to received unreported income.

A bad-debt deduction may not be claimed on any item which has not previously been treated as income or in which the farmer does not have a capital investment. This tends to restrict the bad-debt deduction to accrual-basis taxpayers. A business bad debt is fully deductible. A nonbusiness bad debt is deductible only as a short-term capital loss.

However, a bad-debt deduction by a cash-basis taxpayer is possible. For example, in connection with a business deal, the farmer gives change in cash for a check which is returned unpaid because of insufficient funds, or a business loan goes sour and the obligations become worthless, or a check received in one year is reported as income and not found to be worthless until the next year. The simple failure to collect an account due to a cash-basis taxpayer does not create a bad-debt deduction.

Business bad debts are deductible in the year the debt becomes worthless or partially worthless if the debt is recoverable only in part. It is not necessary that the taxpayer still be in business in order that a debt acquired in connection with the business be a "business" debt. The debt must have arisen in the farmer's business. However, in the hands of a donee, executor or assignee, the debt would be a nonbusiness bad debt deductible only as a short-term capital loss.

A bad debt is a good debt that goes bad. If the farmer knew at the time the debt was created that the borrower could never pay, the deduction will be disallowed.

Trouble often arises on attempts to deduct intra-family loans as bad debts if the creditor fails to make substantial efforts to collect the debt.

Losses suffered as a result of insolvency or dissolution of a cooperative are determined by the nature of the document making allocations to patrons and the relationship created by the bylaws. The patron is treated as having received the face amount of the allocation and either as having reinvested the money in the capital of the association or loaned the money to it. If reinvested, the loss is a capital loss. If a loan was made, a bad debt results.

Losses on loans to corporations are almost always treated as non-business bad debts which are treated as short-term capital losses subject to the limitations on deductibility. Once in a while a taxpayer is able to show that a loan to a supplier or to a business with which he does business is a business bad debt on the theory that it was incurred in a business context rather than as an investment, but the proof is difficult.

Before 1976, an individual guarantor was entitled to a business bad-debt deduction for payment of his obligation as a guarantor, endorser or indemnitor of a noncorporate obligation if the proceeds of the loan were used in the borrower's trade or business and the obligation of the borrower is worthless. After 1975, such payments will be non-business bad debts.

Farm Income Tax Manual, § 423.

§ 4.21 NONDEDUCTIBLE EXPENSES AND LOSSES

Theoretically, all business expenses and losses are deductible, but some capital expenditures are deductible only through depreciation, some costs are never recovered except on sale of property, and some losses between related parties are disallowed.

The broad, general statement that a farmer who operates a farm for profit is entitled to deduct from gross income as necessary expenses all amounts actually expended in the carrying on of the business of farming is not quite accurate.

Farm machinery, buildings, fences, horses, cows, hogs, and other property, may be bought by the farmer to "carry on his business" during the year, but the cost of such items may not be deducted in the year of purchase. Cost of some purchases may be deducted only if spread over the useful life of the asset and claimed in the form of depreciation. Other assets such as feeder cattle and hogs purchased, are neither depreciable nor may the cost be deducted in the year of purchase, but the cost may be deducted in the year of sale.

Capital Expenditures

Except for soil and water conservation expenses, land-clearing expenses and costs of fertilizer and lime, capital expenditures on farm land where depreciable materials are not used result in no credit against gross income until the farm is sold. These expenditures can be used by the taxpayer only to adjust the basis of the farm for the purpose of computing gain or loss on subsequent sale or exchange. At the time of a sale or exchange the costs which were never deducted in prior years are considered a part of the cost of the land.

Land is not depreciable. For example, neither depreciation nor depletion was allowed to a farmer who showed that his land would sink to a point where it would have to be abandoned in fifty years. Depletion was not allowed because there was no mining. Depreciation did not apply because the property was land. He would have a loss upon sale or abandonment. But, there was a case in which the taxpayer was allowed a deduction for peat mud which oxidized and subsided at the rate of 1.1 inches per year.

The cost of acquiring land over which ran irrigation canals carrying water necessary to a farmer's rice fields, which assured him of a continuing water right of way, was a capital expenditure for an asset with an indefinite use which could not be depreciated.

The cost of reforestation is a capital expenditure.

Showing Exhaustion

The problem of the taxpayer seeking a current deduction for depreciation or some form of amortization is to show that the expenditure is exhausted or consumed over a period of time. For example, permanent pasture planted from seed which had to be replanted every ten years was an amortizable asset, while natural pasture was not.

Forest Service grazing permits and state grazing leases have been held not depreciable or amortizable over the remaining period by a taxpayer who purchased a ranch with unexpired leases and permits because there was substantial certainty that the leases and permits would be renewed indefinitely.

Cost of a survey to establish boundaries, ascertain topography, and fix the location of valuable trees and shrubs is a capital expenditure.

Conservation and Land-Clearing Expenses

Unless the special elections to deduct soil and water conservation and land-clearing expenses are made, there is no way for the farmer to claim either as deductible business expenses or depreciation (or amortization) costs of: (1) clearing brush, trees and stumps; (2) leveling and conditioning land; (3) digging ditches; (4) straightening creek beds to prevent erosion; (5) drilling wells which do not have measurable useful lives; and (6) building earthen dams.

On all sales of farm property great care should be exercised to be sure that any such expenditures not deducted are added to the basis of the farm to decrease the farmer's profit (or add to his loss).

Legal Fees

Legal and other fees for the acquisition or defense of title are capital expenditures added to the basis of the property. In addition to fees on the original purchase, this includes, for example, fees incurred in establishing the amount of the dower right in a decedent's estate. Usually there is no definite life over which to depreciate or amortize such costs.

However, if the action involves an accounting or efforts to collect income, it is sometimes possible to deduct the part of the expenses allocable to the production of income or the maintenance of income-producing property.

Expenses and Losses Between Related Taxpayers

A loss on the sale or exchange of property between an individual and a member of his family is not deductible. However, the original transferee of property may deduct his transferor's loss from any gains he might realize.

> **Example:** A farmer sells property to his son for $40,000. The property has a value of $40,000 and a basis of $50,000. The loss is disallowed. Later the market improves and the son sells the property for $60,000. The son's gain is $20,000, but only $10,000 is recognized because he is allowed to deduct the $10,000 loss disallowed to his father.

There is no question of good faith involved. Losses between these related taxpayers are not deductible, honest or not; any taxpayer and his brothers and sisters, spouse, ancestors, lineal descendants; a corportation and an individual who owns more than 50% of the stock; trusts, grantors and beneficiaries; and some others. The contructive stock-ownership rules treat an individual as if he really owned the corporate stock owned by his spouse, family, partners, trusts, other corporations, or related parties. See § 8.4 for similar rules on partners and partnerships.

The disallowance rules apply even though the sale and purchase are made in an open market. For example, if husband sells stock on an exchange at a loss and wife buys similar stock, the loss is disallowed. Even where the husband's property was seized by the Internal Revenue Service for back taxes, and sold at public auction where the wife bid it in, the husband's loss was disallowed.

Unpaid expenses such as salaries or interest due a related person (as defined above) are not deductible by an accrual-basis taxpayer if the expenses are not paid within two and one-half months after the close of the tax year, and the related person is on the cash basis and not required to include these amounts in income.

No Capital Gains

Capital gain is converted into ordinary income on a sale of depreciable property between spouses, between a corporation and a stockholder who, with his spouse and minor descendants, owns more than 80% of the stock, or between two such controlled corporations. Similar rules apply to transactions between a partnership and an 80% partner.

Check List of Farmers' Nondeductible Expenses and Losses

Below are listed expense items which cannot be currently deducted by a farmer:

(1) Buildings, machinery, equipment, livestock, etc., used in the business (except through depreciation).

(2) Crops paid to landlord as rent (cost of raising has already been deducted as an expense).

(3) Expenses not actually paid to a related taxpayer.

(4) Preparation costs which cannot be considered soil or water conservation or land-clearing expenses.

(5) Growing crops sold with the land (expenses must be capitalized).

(6) Labor hired for domestic work.

(7) Livestock in the year of purchase (costs are deducted in the year of sale).

(8) Losses on sales between husband and wife or other related taxpayers.

(9) Losses by "hobby" farmers.

(10) Farm crops and produce fed to livestock or employees (cost of producing these items has already been deducted).

(11) Destruction of growing or harvested crops (costs of producing the crop have been deducted).

(12) Destruction or loss of raised livestock (costs of raising have been deducted).

(13) Shrinkage or deterioration of stored grain (this is reflected in less income when sold or fed to livestock).

(14) Repairs and expenses on the home, household goods or other assets not used in the farming business or held for the production of income.

(15) Transportation of livestock and feed costs en route to the farmer (cost should be capitalized).

(16) Value of the farmer's services for farm labor or building barns, fences, etc.

(17) Value of services of members of family (unless actually paid and treated as income to them).

Farm Income Tax Manual, § 432.

§ 4.22 THE "HOBBY" FARM

A "farm" for tax purposes is a farm in the usually accepted sense. A "farmer" is any individual or organization running a farm for profit. However, persons operating farms for pleasure or

recreation and not for commercial purposes are not "farmers" and their "farm" losses are not deductible.

Deductions for any activity not engaged in for profit, including farms, are subject to special limitations.

For general tax purposes, the term "farm" is used in its ordinarily accepted sense. Stock, dairy, poultry, fruit and truck farms, plantations and ranches are included, as well as any land used for farming operations. For example, the operation of a nursery which specializes in the growing of ornamental plants is a farm.

A "farmer" is an individual, partnership, or corporation cultivating, operating, or managing a farm for gain or profit, either as owner or tenant.

A farm operated for pleasure or recreation and not as a commercial enterprise is regarded as a hobby or personal activity of the taxpayer. The Internal Revenue Service has long permitted the deduction of expenses if a profit actually occurred. If expenses exceeded receipts, both receipts and expenses were ignored.

Tax Reform Act of 1969

For many years, there have been arguments between the Service and farmers who deduct farm losses against other income. The Tax Reform Act of 1969 attacked the problem of "farm loss" deductions through the "excess deductions account," and the recapture of land expenses. In addition, the Congress added a new provision that specifically denied deductions for any activity not engaged in for profit. The statute describes such an activity as one for which no business or nonbusiness deductions would be allowed. The definition is so broad that it will be necessary to use the cases that have arisen in the past to interpret the statute; thus all of the prior learning and experience still has meaning and importance.

The 1969 provision applied specifically to individuals and Subchapter S corporations and it was effective with respect to years beginning after 1969. Thus, the case law and the prior Government policy apply to squabbles for years before 1970 and apparently to arguments about the farm activities of a corporation, an estate, or a trust.

However, the statute has two aspects of importance. First, it allows all personal deductions which would be permitted to a taxpayer in any event, such as interest, taxes, and casualty losses and, in addition, deductions for other expenses up to the remaining amount of gross income. Second, it establishes a presumption in favor of the farmer that he is engaged in a profit-making activity if the income exceeds the deductions in two or more of the five consecutive years ending with the taxable year. In this case, the burden is on the Government to show that the farmer was not really in a business activity. If the activity consists mainly of breeding, showing, training, or racing horses, the presumption is in favor of the taxpayer if he showed a profit in two out of seven years. Failure to show profits in two out of five or seven years, does not mean that the farmer is not engaged in a profit-making activity; it means that he does not get the benefit of the doubt and must prove that he had a reasonable anticipation of putting his operation on a profit-making basis.

The Internal Revenue Service View

Regulations list factors which the Service thinks important in deciding whether an activity is carried on for profit. These factors are:

1. Manner in which the taxpayer carries on the activity. Does he go about it in a business-like fashion keeping books and records and does he operate his business as similar businesses are operated?

2. The expertise of the taxpayer and his advisors. Where the taxpayer has no expertise, does he seek it, and does he follow it once found?

3. The time and effort devoted to the farm. Are either the efforts of the taxpayer or his qualified assistants consistent with the size of the investment and a profit motive?

4. Are the assets expected to appreciate in value? Lack of current income may be offset by the possibility of asset appreciation.

5. The taxpayer's history of profits and losses with respect to the activity. A long uninterrupted history of losses will be harmful.

6. If the taxpayer has been successful in other activities, that may be helpful.

7. If profits are generated, their size in relation to prior or later losses will be considered.

8. If the taxpayer has sufficient wealth to suffer losses, that may indicate a nonprofit motive.

9. If the taxpayer has personal motives or engages in the activity for recreational purposes, his overall profit motives will be suspect.

These regulations also contain several hypothetical fact patterns which may be helpful. See Regs. § 1.183-2(c).

The Courts' View

The preceding paragraphs set out the IRS views on factors bearing on an intention to make a profit. The same question has been considered by the courts, and a few summaries of their views follow.

Continued loss operations are not conclusive against the taxpayer, but he must prove that he had an ultimate hope of profit and that his current and continued losses were necessary to the ultimate profit. However, small receipts and high expenditures tend to suggest the remoteness of profit possibility.

The tendency of consistent losses to imply a nonbusiness operation has been offset by proof of inexperience and unforeseen calamity. Experience of the farm owner and employent of qualified personnel help to show a profit motive. In a number of cases, courts were impressed by the physical work or management activities of the farm owners.

The courts often recognize the long pull in developing breeding lines. A surgeon horse breeder who took in $1900 in three years and paid out $25,000, used professional advice, and showed potential value in his horses, won his case. But failure to cull poor animals and a blind reliance on natural increase to build a herd have lost the deductions.

Generally, a farmer must meet two conditions to be in "business." He must have a motive and an intention of making a profit and his activities must be of a type or character which constitute the present carrying on of a business as distinguished from activities which are preparatory to entering into and carrying on a business at some future date.

Absence of the "country estate" or "show place" aspect and lack of facilities for recreation and entertainment help to prove a profit motive. Separate books for the farm, competent personnel, use of professional farm agencies, subscriptions to farm publications and the like indicate that the farm owner intended to run a business-like operation.

Absence of Profit Motive

The Government will continue to challenge the losses of urban-based farmers on the theory that the farm is really a hobby. It is important to be conscious of the extra measure of proof needed to show that the farm owner who has another livelihood is operating his farm with a profit-making motive. The challenge from revenue agents comes when the farm owner does not fit the picture of a local farmer.

A large residence on a farm suggests that it is used primarily for personal purposes. The residential use of the farm by a taxpayer who is not a full-time farmer weighs heavily against him, particularly if the residence is out of proportion to the operation. The fact that the taxpayer enjoys living on the farm does not prove that he is not in business; but it certainly does not help to prove that he is in business. The tip-off on residential farms is small acreage and high taxes. Swimming pools, barbecue pits, and extensive patios are suspect.

Sometimes a taxpayer claims that he has converted his hobby, for example, breeding and showing of livestock, to a business operation but he continues to suffer losses. The Government has argued that the taxpayer would have continued his hobby activities whether he made a profit or not and, therefore, his true motive was not profit seeking.

The Government has also made the point that a farm was too small for economic operations; for example, it argued that a forty-acre farm was not large enough to be operated profitably and that, therefore, the taxpayer could not have a profit motive. In another case the Government claimed that the bad record of losses in prior years would have led a sensible man to abandon the operation; therefore, it could not now be regarded as a potential profit-maker.

In the early stages of an operation, the Government has challenged the deduction of expenses on the theory that the farm is not yet a business and that expenses must be capitalized until the business is in full fledged operations. See § 4.12.

The corporate farm is suspect if it serves as a vehicle for stockholder enjoyment of a fancy farm, a residence or a hobby, or if transactions between stockholders and corporation lack substance. Any arrangements between the controlling stockholders and the corporation that veer very far from the standard patterns of the particular area or the fair rental value of the premises will be questioned.

All of these arguments are still valid under current law. The amendments of 1969 and 1976 do not set up any new rules for determination of the existence of a profit-making motive except for the presumption which helps the farmer in his burden of proof. The same old theories of profit making will be used by farmer and Government.

Farmer or Investor

The Service seems to have a notion that there is a difference between a farmer and an investor in farms. The theory leads to the conclusion that an investor is not entitled to use farmer rules, for example, the right to use cash-basis accounting, and that an investor should capitalize the costs of raising animals to maturity instead of deducting them. Such a distinction seems highly unsound but its possible use by agents cannot be ignored.

Even the Congress, faced by the claim that farms were serving as "tax shelters" for wealthy taxpayers, accepted this dichotomy in enacting the "excess deductions account" in 1969. A farmer did not have to maintain an EDA unless he had over $25,000 of farm losses for the year and also had $50,000 of nonfarm income. See § 5.13. The rules changed in 1976. No additions are to be made to the excess deductions account after 1975. The 1976 Act changes purport to distinguish between the passive investor and the real farmer. The "at risk" rule and the farm syndicate rules try to distinguish the real "dirt" farmer from the seeker after farm tax shelters. See § 4.1.

In a recent case, an appeals court on the East Coast went out of its way to deplore what it regarded as a developing tendency in the lower courts to emphasize the fact of the taxpayer's wealth, his love of farming, and the losses he incurs as indicative of an absence of a profit motive. The court suggests that the true factors to be considered are whether the expectation of ultimate profit is reasonable, how the operation is conducted, to what extent the taxpayer personally engages in it, and the likehood that the activity is within the normal activities of people who have some degree of wealth. The appellate court seemed to think that the lower courts' approach to the motives of the rich was quite different than the approach to the motives of the less affluent participators in farming ventures. As was noted above, this is one factor tested by regulations under the hobby loss provisions.

§ 4.23 CREDIT AGAINST TAX FOR WIN LABOR

A welfare recipients' work incentive program (WIN) authorizes an income tax credit for workers hired pursuant to the program.

Advantages and Procedures

Employers may claim a dollar-for-dollar credit on their income tax for 20% of wages paid to each WIN employee and each former welfare recipient for the first twelve months of the employee's employment—consecutive or not—provided that the employment continues at least 180 days. No credit is allowed for wages after 24 months from the first day of employment.

The credit is recaptured if employment is terminated without cause within the first 180 days of employment. A discharge due to lack of business is the equivalent of cause.

Other restrictions are:

The WIN employee must work in the United States; cannot be a close relative of the employer; must be paid the same wages as others doing comparable work; and must work in the employer's normal trade or business.

The employer cannot receive a credit for wages reimbursed under other programs; cannot receive a credit in excess of his income tax; cannot receive full credit in excess of $50,000 (amounts in excess of $50,000 receive one-half credit); must refund credits received if the employee is terminated during the first 180 days without cause; must provide minimum safety standards and benefit provisions.

All WIN employees must be certified by the State Employment Office; such certification will not be granted if another employee has been displaced to hire the WIN employee.

The credit for welfare recipients is to terminate at the end of 1979.

There is also a credit for wages paid to former welfare employees on wages up to $5000 even though the wages are not deductible.

Farm Income Tax Manual, § 101.

§ 4.24 CONTRIBUTIONS TO RETIREMENT PLANS

Contributions to retirement plans may be deducted. The amounts of the deductions are limited and depend upon the way the business is organized and the kind of plan chosen.

The Employee's Retirement Income Security Act of 1974 (ERISA) added complex rules on deferred compensation. The details of the available plans are beyond the scope of this book, but the outline of each kind of plan can be suggested. The tax savings are significant, and most farmers and ranchers should talk to a tax adviser about a plan.

There are three kinds of retirement plans: (1) plans for the self-employed, including partnerships, (2) plans for corporations, but Subchapter S corporations are subject to some of the rules for self-employed plans, and (3) individual retirement accounts.

(a) Self-employed plans

A self-employed plan may be adopted by any person subject to the self-employment tax and by ministers and Christian Science practitioners. A partnership may adopt a plan for its partners. An "owner-employee," i.e., a sole proprietor or a partner owning 10% of the capital or profits of the partnership, may not participate in the plan unless all employees who have three years of service amounting to 1000 hours of work each year are also covered. A spouse who is a full-time employee must be covered even though he or she is not included under Social Security. Employees of other businesses controlled by the employee-owners must be included in the plan. Control means ownership of an unincorporated business or 50% ownership in partnership capital and profits by one or more owner-employees.

Funding of a retirement plan may be accomplished by (a) contributions to a trust of which a bank is trustee; (b) purchase of life insurance, annuity or endowment contracts from an insurance company, with or without a trustee or custodian; (c) purchase of shares of a mutual fund or a regulated investment company to be held by a bank as custodian; (d) purchase of special Government bonds the terms of which read like a retirement plan, with or without a trustee or custodian; or (e) purchase of "face amount certificates" which are nontransferable if owned by a person other than a qualified trust. IRC Sec. 401.

Except for certain defined benefit plans, contributions for owner-employees are limited to 15% of earned income or $7500, whichever is lower. In any event, an owner-employee may contribute and deduct the lower of $750 or all of his earned income provided that his adjusted gross income does not exceed $15,000.

All self-employment income from farming constitutes earned income as long as services by the owner are a material income-producing factor. If a self-employed individual devotes substantially all his time to the farm operation, his personal services are treated as a material income-producing factor in that business. If less than full-time effort is devoted to the business, the respective contributions made by personal services and capital are considered in deciding whether personal services are a material income-producing factor.

Other Terms

There may be no discrimination in benefits for top employees. Contributions for the benefit of employees must vest immediately if a plan involves "owner-employees." There can be no forfeiture of an employee's rights if he leaves his employment.

Benefits to "owner-employees" may not begin before age 59 1/2 except upon permanent disability or death, and this restriction must be written into the plan. Payments must commence no later than 70 1/2 for an "owner-employee" and at age 70 1/2 or the year of retirement, whichever is later, for others. Generally a distribution on death of an "owner-employee" must be completed within five years or five years from his spouse's subsequent death, or converted into an immediate annuity.

Plans may be coordinated with Social Security if the contributions for "owner-employees" are not more than one-third of the total. Rules differ from those for integration under regular employee plans.

Generally, the funds set aside may not be used for any purpose until retirement. Penalties are provided for premature distributions or improper use of assets or for improper excess contributions for an "owner-employee." Transactions between the self-employed individual and the plan are generally forbidden.

A plan may be either a pension plan or a profit-sharing plan. A pension plan requires the employer to contribute either an amount necessary for a fixed benefit at retirement (a "defined benefit") or a money purchase contribution based on annual compensation, years of service and other factors. Profit-sharing-type plans require a contribution of a predetermined portion of the operation's profits. Contributions for owner-employees are subject to the limitation of $7500 and 15% of income. Due to these limitations, a pension plan, which generally will require fairly stable contributions for common-law employees, may result in substantial contributions for persons who are not the owners and a very low contribution for the owners in years in which the earned income is low. Since farm income may fluctuate, a farmer may find that a profit-sharing plan, in which all contributions are based upon farm profits, is better than a pension plan even though in some years the contributions will be less than the maximum amount which might have been contributed for the owner-employee under a pension-type plan.

The contribution of a fixed percentage of earned income (regardless of cash available) allocated on the basis of relative income will not discriminate in favor of owner-employees. Thus, contributions made for employees may be the same percentage of their compensation as the percentage of earned income contributed for owner-employees. For example, if the earned income of a sole proprietor is $50,000, he could contribute $7500 on his own behalf and, if a formula based on compensation were used, he would be required to contribute 15% of the compensation of his employees.

Excess Contributions

Plans may provide for employee contributions in addition to the deductible employer contribution. The "owner-employee" may make excess nondeductible contributions equal to the greater of 10% of earned income or $2500 for himself so long as employees may contribute at the same rate. While there is no deduction for the extra amounts, the earnings thereon are not taxed until distributed. If there are no employees, no extra contribution can be made by a self-employed person. An excise tax of 6% is imposed on contributions in excess of these limits.

Distributions

Distributions are taxed in two ways. If the distribution is received as an annuity, the general annuity rules of Section 72 apply. If a lump-sum distribution is made, a part of it may be treated as capital gain, and the balance is subject to a ten-year forward averaging provision. The capital gain portion is the fractional part established by a fraction with years before 1974 as the numerator and total years in the plan as the denominator. This gain may be averaged under the regular averaging provision. A taxpayer may now elect to treat pre-1974 contributions as post-1973 contributions and treat them as ordinary income subject to the ten-year forward averaging provision. A lump-sum distribution is payment of the employee's account within one taxable year on account of (1) death, (2) reaching age 59 1/2,

(3) separation from service, or (4) disability. The distribution must be in one taxable year but not necessarily when the event occurred.

(b) Corporate plans

Additional tax benefits are available to corporations other than Subchapter S corporations. They are higher annual contributions, a greater ability to avoid contributions on behalf of employees who are not equity owners, and vesting requirements that may result in forfeiture of benefits for employees who terminate.

The limitations on maximum contributions are 25% of compensation or $25,000 except profit-sharing plans are limited to 15% of compensation. This $25,000 is also subject to revision for changes in the cost of living.

As with self-employed plans, there are rules against discrimination in favor of high salaried employees. However, the rules on integration with social security are much more liberal. The maximum contribution on behalf of owner-employees must be reduced by the self-employment tax.

Vesting

Vesting may be deferred as long as ten years. Unvested amounts are forfeited and may benefit the other beneficiaries. But even though benefits are not vested, all employees who have completed one year of service and who are 25 years old must be covered unless the plan provides for vesting for all employees who are 25 or more.

There are other complex compliance requirements that must be followed by a corporation to maintain qualification of its plan. Included among these requirements are reports which must be filed with the Labor and Treasury Departments, fiduciary requirements, and accounting and actuarial reports. The operational requirements, in addition to the costs of implementing a qualified plan, will in most cases substantially exceed the costs under a master or prototype plan for the self-employed.

Distributions

With one favorable exception, distributions are subject to the same rules as those from self-employed plans. Under the exception even the capital gains tax is not paid on appreciation in value of the employee's securities included in the lump-sum distribution.

(c) Individual retirement accounts

Since enactment of ERISA in 1974, both employees and self-employed may establish an individual retirement account if not covered by a qualified plan. The deduction is the lesser of $1500 or 15% of compensation. The 1976 Act enlarges this deduction slightly where contributions are made each spouse. The limit for both spouses is the lesser of 15% of earned income or $1750 ($875 for each spouse), and the contribution must be to separate accounts or one account with a subaccount for each spouse. There is no deduction for contributions to a jointly owned account. There may be a right-of-survivorship between the two accounts or the two subaccounts where only one account is sued. A taxpayer must choose between the making of contributions up to $1500 for himself or making contributions for both spouses. Only one course is open to him. Form 5329 must be filed with the taxpayer's return.

Distributions

Withdrawals (other than annuity policies) from the account are treated as taxable income. Payments from an annuity policy fall under Section 72, and the policy is treated as having a zero basis. Withdrawals may not begin before age 59 1/2, and they must commence before age 70 1/2.

For estate tax treatment, see § 9.4.

chapter 5

TAX MANAGEMENT OF LAND AND DEPRECIABLE PROPERTY

Section

§ 5.1 INTRODUCTION

The major technique for the recovery of a farmer's investment in depreciable property is the allowance of a deduction for depreciation, including some special provisions for rapid write-off.

An investment credit against the income tax is allowed for most depreciable property other than buildings.

Upon the sale or disposition of an asset upon which depreciation or an investment credit has been claimed, the farmer must return

to income excessive depreciation and add back to the tax any "unearned" portion of investment credit previously claimed.

See the quick reference charts summarizing rules on depreciable property and recapture at the end of this chapter.

This chapter deals with various facets of the handling of depreciable property. Ordinary depreciation has always been allowed as a deduction for income tax purposes. More recently, there were added fast depreciation methods, extra 20% depreciation, the investment credit, special rules on salvage value, the provisions for recapture of depreciation, and the Asset Depreciation Range (ADR) system.

The methods of fast write-off brought additional problems. Depreciation is deducted against ordinary income, but most depreciable property falls under Section 1231 of the Internal Revenue Code and usually produces a capital gain on sale. If property subject to depreciation is sold at a gain, the practical result is the conversion of ordinary income to capital gain.

In 1962, the Congress enacted Section 1245 of the Internal Revenue Code to tax (i.e., to "recapture") as ordinary income on the sale of depreciable property (except livestock and buildings) so much of the capital gain as represented a recovery of previous depreciation. In 1964, a similar provision was added to recapture some of the depreciation taken on buildings. In 1969, the Congress made post-1969 depreciation on livestock recapturable and extended the amount of recapture on real estate for 1970 and thereafter.

The investment credit is subtracted not from income but from the computed tax and any "unearned" portion of the credit is added back to the tax in the year of sale of an asset upon which the credit was claimed.

In 1966, the Congress suspended the allowance of the investment credit from October 10, 1966 to the end of 1967 because the economy seemed to need some cooling off. But in 1967 the economy apparently had cooled off too much and the investment credit was restored effective March 10, 1967. The Tax Reform Act of 1969 repealed the investment credit as of April 19, 1969 but left in effect the recapture provisions for investment credits claimed before the repeal. But again in 1971, the economy seemed to need stimulus. The credit was restored for orders after March 31, 1971, and acquisitions after August 15, 1971. It is subject to recapture.

Congress banned the use of the fast methods of depreciation on real estate. In 1969, depreciation on new buildings (except residential rentals) was limited to the 150% declining balance and on used buildings to straight line.

Recapture of deductible land expenses (soil and water conservation expenditures and land-clearing expenses) claimed after 1969 was added in 1969. Another new concept, the recapture of farm losses against capital gains by farmers with nonfarm income, was also enacted in 1969, but farm losses after 1975 will not be recaptured.

To understand the problems of land and depreciable property, it is necessary to examine not only the techniques of recovery of the farmer's cost but also the possible recapture of his previously claimed deductions or credits. It should be noted that although these provisions are often based upon the same general policies, the rules and details may be quite different.

The tax aspects of depreciable property can best be appreciated by examining the "big picture"—an over-all view of all of the different rules with which a farmer must deal.

The **quick reference charts** at the end of this chapter offer summaries of the whole complex of depreciable property and recapture provisions and show the interrelationship of the various rules that must be understood.

This chapter follows roughly the order of a farmer's practical application of the provisions relating to depreciable property, the extra first-year depreciation, ordinary depreciation and the so-called "fast" depreciation. Subsequent sections discuss more general background aspects: property which is depreciable, salvage value, useful life, basis for depreciation. The final sections deal with investment credit, investment credit recapture, recapture of depreciation on personal and real property, and the provisions on recapture of land expenses and farm losses in the "excess deductions account."

A short section reviews the aspects of depletion of natural resources that might involve a farmer.

Farm Income Tax Manual, §§ 500, 501.

§ 5.2 EXTRA FIRST-YEAR DEPRECIATION

Tangible personal property with a useful life of at least six years may have an extra depreciation allowance in the year of acquisition. At the taxpayer's election, he may deduct up to 20% of the cost of the property. Regular depreciation is allowed on the remaining cost.

As an additional fast write-off of part of the cost of depreciable property, a farmer may deduct up to 20% of the cost of an asset in the year of purchase. The remainder of the cost is depreciated in the usual manner.

If a taxpayer elects as additional first-year depreciation a lesser portion of the cost of qualifying property than the full 20%, he cannot increase the deduction to the full 20% allowable after the due date of the return.

The election applies only to tangible personal property of a character subject to the allowance for depreciation purchased for use in the trade or business or to be held for the production of income. The property may be new or used but it must have a useful life of at least six years.

Buildings, structures and improvements to real property are not within the election. Farm equipment, machinery, movable structures that do not become part of the real estate, and draft, breeding, dairy and sporting livestock qualify. But the Internal Revenue Service disallows the extra depreciation on prefabricated grain bins and similar structures which are inherently permanent even though they may be taken apart and reassembled at other locations.

Orchards and groves are not tangible personal property and therefore are not eligible for the extra 20% depreciation.

Purchased or Exchanged Property

The property must have been acquired by purchase or exchange. The extra depreciation allowance does not apply to property acquired by gift, bequest or inheritance. If the property is acquired in a tax-free exchange, the 20% write-off applies only to the boot paid on the exchange.

Property acquired from a related seller is not eligible. Property purchased from a person or organization whose relationship to the buyer is such that losses between them would be disallowed cannot qualify for the write-off. See § 4.21. Purchases from brothers or sisters are permissible.

Individuals, partnerships, corporations and estates, but not trusts, may use the deduction.

Computation of Deduction

In the year of acquisition of the property, the taxpayer may take a deduction up to 20% of the cost of the asset, plus the regular depreciation allowed for the first year or part of the year. Salvage value is not considered in computing the 20% allowance even though it is required for regular depreciation. The basis of the property subject to the election is reduced by the amount of the extra deduction claimed and the remaining cost of the property is subject to regular depreciation.

Example: If a farmer on a calendar year purchased an asset for $5000 on July 1 and he elected to take the 20% write-off, he would deduct $1000. If the asset had a six-year life, a salvage value of $400 and was depreciated on the straight-line method, he would deduct a half year's regular depreciation. This would be $300. ($5000 cost less $1000 extra depreciation leaves a basis for regular depreciation of $4000. Reduced by salvage value, the depreciation basis is $3600 and depreciation is $600 per year. Half of a year's depreciation is $300.)

Trade-Ins

The 20% depreciation applies to property received on a trade-in only to the extent of the cash paid on the trade-in. Sale of the old property and purchase of new rather than trading will be necessary if the full deduction is to be used. However, see the warning on phony trades at § 3.27.

When Deducted

The extra deduction is to be taken in the first taxable year for which the taxpayer is entitled to depreciate the property. Normally, this will involve no problem. However, if the farmer has a consistent practice of holding items bought late in the current year for the next year's depreciation schedule, he should follow the same pattern in electing the 20% write-off.

If depreciable livestock is purchased when immature, some authorities suggest that depreciation is not taken until the animal is mature. In the year of maturity, both the 20% depreciation and regular depreciation would be claimed.

The election is made by claiming the deduction on the tax return for the year of acquisition. The election relates to each piece of property. Once made, the election is irrevocable but only for the particular asset. Otherwise, it is made for each particular asset in the year of acquisition.

Maximum Depreciation in Year

Although the taxpayer makes his election each year for property acquired in that year, he is limited to $10,000 worth of property per year on a separate return ($2000 deduction) and to $20,000 worth of property per year on a joint return ($4000 deduction). The taxpayer selects the cost (or fractional part of the cost) of one or more of the assets acquired during the year.

A partnership makes the election, but each partner claims his share of the partnership's deduction which, in years after 1975, may not exceed $2000. Before 1976, the partnership could pass through as much extra depreciation as each partner needed to fill his personal limit of $2000 or $4000 on a joint return. Subchapter S corporations, partnerships and estates are limited to a $2000 deduction on $10,000 of property.

Farm Income Tax Manual, § 503.

§ 5.3 ORDINARY DEPRECIATION

The cost of a physical asset that has a determinable useful life of more than a year is recovered through an allowance for depreciation. Cost of intangible assets is recovered through amortization.

Depreciation is the exhaustion, wear and tear or using up of an asset used in the farm business. The cost of the asset is recovered as a deduction against income over the years of the use of the asset. The allowance for depreciation is the amount that should be set aside for the taxable year in accordance with a reasonably consistent plan so that the aggregate of the amounts set aside, plus the salvage value, will at the end of the useful life of the asset equal the cost or other basis. See § 5.5 for the methods of depreciation.

If an intangible asset, such as a leasehold or a patent, has a useful life, the taxpayer's investment is recovered in similar manner by amortizing the cost over the useful life. Intangible amortizable assets do not qualify for extra first-year depreciation or fast depreciation.

If the owner of an asset does not suffer any depreciation, he is not entitled to a deduction. For example, in the case of Kem, Jr., 51 TC 44 (1968), the court disallowed depreciation to the lessor of a herd of breeding cows because the lease provided for the lessee to cull a specified percentage of the herd each year and replace them with young animals so as to maintain a herd of comparable quality and age as that originally leased. The case was upheld on appeal.

When Depreciation Is Claimed

Depreciation begins when an asset is placed in service, that is, when it is acquired and held for use. A building is depreciable when construction is completed. Equipment is usually depreciable upon acquisition. The fact that an asset is held idle for a time does not preclude depreciation. Additions to structures or costs of reconditioning equipment become additions to the depreciation schedule upon completion. No depreciation is permitted during construction. Depreciation begins on purchased property when the benefits and burdens of ownership pass to the buyer, even though title is delayed.

Livestock held for draft, breeding, dairy, or sporting purposes is depreciable but practice varies on when the depreciation should start.

Some authorities take the position that the depreciable useful life of the animal cannot begin before maturity.

Depreciation must be claimed when it occurs. A taxpayer may not take advantage in later years of his prior failure to deduct the depreciation allowance or of his action in deducting an allowance plainly inadequate under the known facts in prior years.

An asset acquired during the year, as most of them are, is properly depreciated only for the period of the year that it was owned. However, the "half-year convention" (which deems all acquisitions to have been made on the first day of the second half of the year so as to yield six months depreciation on all acquisitions) may be used. If the ADR system is elected, the "modified half-year convention" may be used. Under it all acquisitions in the first half of the year are deemed held a full year, and all acquisitions in the second half of the year are first depreciated at the beginning of the following year. See § 5.6.

Form 4562 may be used to report depreciation. It is an optional attachment to return forms for individuals, partnerships, fiduciaries, or corporations. If ADR or its equivalent for pre-1971 acquisitions (the Guideline Class System) is elected, Form 4832 (ADR) or Form 5006 (Guideline Class System) must be used.

Obsolescence

Normal obsolescence is included in depreciation. An addition to depreciation can be made for obsolescence where the expected life is shortened by technological development or economic changes.

Abnormal or extraordinary obsolescence is generally caused by rapid advances in technology or by sudden economic changes which could not reasonably have been foreseen when the taxpayer acquired the asset. When the obsolescence becomes clear or discernible, the useful life of the asset should be adjusted and the obsolescence taken as part of the depreciation deduction over the remaining new useful life. For example, if a special building or a confined hog, poultry or dairy cow unit was becoming technically obsolete but not abandoned, the useful life should be shortened and the depreciation stepped up.

An abandonment loss results only from a sudden or unforeseen loss which completely terminates the useful value of the asset. An abandonment loss may not be claimed if the useful life of the asset terminates solely as a result of the gradual processes for which depreciation allowances are authorized. Also, if ADR has been elected, there are even greater restrictions on the allowance of a loss. See § 4.19.

Farm Income Tax Manual, §§ 504, 508, 512.

§ 5.4 PROPERTY SUBJECT TO DEPRECIATION

Only property used in the trade or business or held for the production of income is depreciable. Unless the life of the property is determinable, no depreciation may be claimed.

Only persons who have a capital or proprietary interest in depreciable property are entitled to claim the deduction.

Depreciation is allowed only upon tangible property used in the trade or business or held for the production of income. There is no deduction for depreciation on inventories or stock in trade or land. Land is regarded as permanent, having no measurable useful life. Intangibles are not depreciable but their cost is recovered through deductions for amortization in annual installments if they have a measurable useful life. Natural resources are subject to depletion, not depreciation. See § 5.14. Depreciation is not allowed on the taxpayer's residence, personally used automobile, or personal effects, since they are not used in the trade or business, nor are they held for the production of income.

What May Be Depreciated

Even though property is used in the trade or business, no depreciation will be allowed unless the asset has a fixed and determinable existence. Exhaustion which may be taking place must be susceptible of measurement. Some of the capital investments made by the farmer which are considered nondepreciable for this reason are: clearing brush, trees and stumps; leveling and conditioning land; digging ditches; straightening creek beds to prevent erosion; building earthen dams; and land costs.

However, many of the expenses which cannot be recovered through depreciation under the above rule qualify as deductions under the elections to deduct soil and water conservation and land-clearing expenses. See §§ 4.13, 4.14.

Generally the grading of land is a part of the cost of the land itself and is therefore not depreciable. However, under some circumstances the cost of digging and removing soil for the proper setting of buildings and the paving of roadways may be considered a part of the cost of the building or the roadway and therefore a part of the depreciable cost of the item.

Depreciation should be claimed on an asset even though it was not used during the year. So long as the asset has not actually been abandoned, it is still regarded as used in the trade or business.

If personal assets are converted to a business use or otherwise become held for the production of income, they then become depreciable.

There has been some feeling that poultry, because it is excluded specifically from the definition of livestock under Section 1231, is not depreciable. Few taxpayers are inclined to depreciate poultry, but if chickens are used for breeding or egg-laying purposes for substantially more than a year, they would be depreciable assets. Very often the cost will have been deducted when the poultry was acquired, and there would not be any basis for depreciation. But if poultry is purchased by a farm syndicate, as defined in the 1976 Act, its cost must be capitalized. See § 4.1.

An interesting illustration of the theory of depreciation involved a garbage collector who bought a hole in the ground in which to bury his garbage. He depreciated his cost over the period of time that it would take to fill the hole. One court thought he made sense, although some judges dissented and the Government objected to what it called depreciation of land. Six years later the Government acquiesced.

Who May Depreciate

Depreciation is ordinarily claimed on tangible property by the owner who uses it in his trade or business or for the production of income. If a person has no proprietary interest in the property, he must have a capital interest and suffer the loss resulting from exhaustion or wear and tear. For example, one who has transferred property as security for a loan may claim depreciation on it; so may the purchaser of property under a contract from the time that possession or benefits and burdens of ownership are transferred to him.

The cost of improvements made by a lessee at his own expense may be amortized over the life of the lease or depreciated over the life of the asset, whichever is shorter. A lessor cannot depreciate assets added to the leased premises by the lessee because the lessor has no investment in them, no cost to recover. However, if the lessor has accepted the improvements in lieu of rent, then he has an investment and is entitled to depreciate the assets using the rent as the basis, just as if he had received the actual rent and made the improvements himself.

Farm Income Tax Manual, §§ 506, 507.

§ 5.5 METHODS OF COMPUTING DEPRECIATION

Methods of depreciation of importance to the farmer are (1) straight-line method; (2) declining-balance method; (3) sum-of-the-digits method; and (4) possibly others, such as the unit-of-production method.

Depreciation methods on real estate acquired after July 24, 1969 are limited to the 150% declining-balance method on new buildings and the straight-line method on used buildings.

Most farmers will establish depreciation allowances using the straight-line method, the declining-balance method or the sum-of-the-digits method. In addition, limited use may be made of the unit-of-production method or other techniques peculiarly adapted to the type of asset or the type of farming.

A method of depreciation is adopted for a piece of property the first time that it appears on the income tax return. Any method so adopted must be continued for that particular piece of property unless the Commissioner authorizes a change. However, change from double declining-balance to straight-line method may be made at any time without permission.

Straight-Line Method

The straight-line method of depreciation is the simplest form of depreciation. The cost basis of the asset (less its estimated salvage value, see § 5.7) is divided into equal annual installments over the property's estimated useful life.

> **Example:** A farmer buys a tractor for $5000 on July 1. It has an estimated useful life to him of eight years and an estimated salvage value of $1000. His depreciation allowance is $4000 ÷ 8 or $500 per year. For the current calendar year, he will claim half of that or $250. If he claimed the 20% first-year depreciation of $1000, the basis for regular depreciation would be $4000 less salvage of $1000 or $3000. Annual depreciation on an eight-year life would be $375 of which $187.50 would be claimed this year.

In computing salvage value a farmer may reduce the true salvage value of an asset with a life of three years or more by 10% of its cost. In the above example, he could ignore $500 of salvage and depreciate $4500, or if he chose the extra 20% first-year depreciation, he would claim it and have $3500 to depreciate over the eight-year life. If salvage value were less than $500, the farmer in the above example would not have to use any salvage value. See § 5.7.

If the farmer had elected the ADR System, he may ignore salvage value in computing his annual deduction. Thus, his cost of $5000 would be spread over the eight-year life to give a deduction of $625 per year until salvage value is reached. If he also chose the first-year extra 20% depreciation, he would have $4000 to spread over the balance of the life to yield an annual deduction of $500 ($4000 ÷ 8 years).

While, in the instances just noted, salvage value may be ignored for the purpose of computing the annual depreciation allowance, no property may be depreciated below its salvage value. Thus, if salvage value were $1000, depreciation would stop when the basis was reduced to that amount by prior depreciation.

Double Declining-Balance Method

The rate of depreciation for the first year under the declining-balance method may be twice that of the straight-line method. This 200% depreciation technique recovers twice as much cost in the first year and a much higher percentage of cost during the early years of the use of the property than the ordinary straight-line method.

The double declining-balance method applies only to tangible property (except a building or its structural components) with a useful life of three years or more. The useful life of the asset is related to the particular taxpayer rather than to the useful life of the asset as such. For example, if a farmer trades his business car every two years, the useful life is two years and the fast depreciation methods would not apply.

The original use of the property must commence with the taxpayer. The words "original use" are severely limited. For example, if a farmer started the double declining-balance method on property and then contributed the property to a partnership, the partnership could not continue the fast method because it was not the original user. The method may not be continued or started on any property in the hands of a distributee, vendee, transferee, donee, or grantee from the original owner.

To use the double declining-balance method the procedure is: (1) determine the rate under the straight-line method; (2) double that rate; and (3) apply the rate against the adjusted balance each year.

> **Example:** A farmer buys an asset for $2000. Estimated useful life is ten years. This means a rate of 10% a year or $200 under the straight-line method.
>
> Under the declining-balance method the rate would be double the straight-line rate (20%) with results as follows:
>
> 1st year—20% of $2000 equals $400 (depreciation allowable)
>
> 2nd year—$2000 less $400 (depreciation previously allowed) equals $1600 (adjusted basis)
> 20% of $1600 equals $320 (depreciation allowable)
>
> 3rd year—$1600 less $320 equals $1280 (adjusted basis)
> 20% of $1280 equals $256 (depreciation allowable)

The declining-balance method permits an immediate high deduction as contrasted with an equal amount each year. For an asset with a useful life of ten years, the farmer will recover over 40% of the cost in the first quarter of the life of the asset, and in the first half of the useful life he will recover about two-thirds of the cost.

A taxpayer may at any time change from the declining-balance method to the straight-line method without permission. The shift to straight-line depreciation must be maintained and cannot be again changed without permission. The declining-balance method leaves an undepreciated residue at the end of the depreciation period. If the asset really has no salvage value, the taxpayer should shift to straight-line depreciation in time to absorb this residue. In no event may an asset be depreciated below a reasonable salvage value even under the double declining-balance method.

The double declining-balance method may not be used on buildings or building improvements (other than residential rental units) acquired or constructed after July 24, 1969. However, if the method was adopted on buildings acquired or constructed before that date, it may be continued. Of course, a farmer can change his method to straight line without permission.

150% Declining-Balance Method

A more limited form of declining-balance method of depreciation also is available. It can be applied to new buildings and their structural components (but not used buildings) or to used properties other than buildings. The difference between the two methods is that under the alternative method, the rate is limited to 1 1/2 times the straight-line rate as contrasted with twice the straight-line rate. The 1 1/2 declining balance may be elected for any piece of property in the return for the first year in which the property is subject to a depreciation allowance except for used assets with less than a three-year life.

Salvage value problems are postponed because any declining-balance method leaves an undepreciated residue. Whether this is proper salvage value cannot arise until the last year or years of life.

Here is how the residues under the 150% and the 200% declining-balance rates would compare with those under the straight-line rate. Assume a $5000 piece of equipment with a five-year life:

	Straight line 20%	150% Decl. Bal. 30%	200% Decl. Bal. 40%
1st year	$4000	$3500.00	$3000.00
2nd year	3000	2450.00	1800.00
3rd year	2000	1715.00	1080.00
4th year	1000	1200.50	648.00
5th year	0	840.35	388.80

A taxpayer may shift to the straight-line method and continue depreciation if the residue under the declining-balance method is in excess of salvage value.

Sum-of-the-Digits Method

The sum-of-the-digits method is another declining form of depreciation which gives higher deductions in the earlier years of useful life. The depreciation allowance is computed by applying a declining fraction to the cost of the property reduced by salvage value. In the fraction, the numerator is the remaining years of useful life and the denominator is the sum of the digits of each of the years of useful life.

Example: A farmer buys an asset for \$2200. The expected useful life is ten years and the salvage value is \$200. He computes his depreciation as follows:

First—Compute the sum of the years' digits:

1
2
3
4
5
6
7
8
9
10

Total 55

A quick way to compute the sum of the digits for the various years of useful life uses a simple formula. Add 1 to the number of years of useful life of the asset and multiply the result by half the number of years of useful life. For example, on an asset with a ten-year life: $(10 + 1) \times 5 = 55$.

Second—Divide the depreciation basis of \$2000 (\$2200 cost minus \$200 salvage value) by 55 (the sum of the digits) to get the lowest unit of depreciation to be allowed (in the tenth year of life) $\$2000 \div 55 = \$36.36$.

Third—Compute the actual depreciation to be allowed each year as follows:

1st year	10 x $36.36	*$364.00	(10/55ths of $2000)
2nd year	9 x $36.36	* 327.00	(9/55ths of $2000)
3rd year	8 x $36.36	* 291.00	(8/55ths of $2000)
4th year	7 x $36.36	* 255.00	(7/55ths of $2000)
5th year	6 x $36.36	* 218.00	(6/55ths of $2000)
6th year	5 x $36.36	* 182.00	(5/55ths of $2000)
7th year	4 x $36.36	* 145.00	(4/55ths of $2000)
8th year	3 x $36.36	* 109.00	(3/55ths of $2000)
9th year	2 x $36.36	* 73.00	(2/55ths of $2000)
10th year	1 x $36.36	* 36.00	(1/55ths of $2000)
Total		$2000.00	

*Final figures are rounded to the nearest dollar as they will probably be reported on the tax return.

Sometimes it is handy to have the salvage value round out the figures for easy computation. If salvage value is $275, the lowest unit of depreciation is an easy $35.

Here again, salvage could be ignored under the 10% rule mentioned above. Similarly, salvage value may be ignored in the above computation if the taxpayer has chosen the ADR System. Thus, instead of dividing $2000 (cost of $2200 less $200 salvage) by 55, the taxpayer would divide $2200 by 55 giving $40 to be multiplied by the appropriate numerator, i.e., 10 in the first year to yield a $400 deduction, etc. In this case, there would be no deduction after the 8th year because the undepreciated balance would be $200, its salvage value, and no asset may be depreciated below salvage value.

The depreciation deduction under the sum-of-the-digits method will not be as great in the first year as under the double declining-balance method but the second-year deduction will be greater under the sum-of-the-digits method. Results at one-fourth and one-half the useful life of the asset will be substantially the same as with the declining-balance method.

One drawback of the sum-of-the-digits method is the problem created in computing depreciation on assets acquired during the year. Unfortunately, farmers do not purchase farm machinery and other assets on January 1 of each year as is often illustrated in sample forms. Consequently, the depreciation schedule frequently shows items owned for fractional parts of a year or at least for half years.

If the first year of the ownership of the asset does not coincide with the tax year of the taxpayer, the first tax year's depreciation will be a portion of the depreciation computed on the first ownership year. In the second tax year, the depreciation will be the rest of the depreciation computed for the first year of ownership and a portion of the depreciation figured for the second ownership year.

Like the declining-balance method, the sum-of-the-digits method applies only to the original user of the asset and only to property having a useful life of three years or more.

It, too, may not be used on buildings (other than residential rental units) and their structural components constructed or acquired after July 24, 1969.

Other Methods

Any other method of depreciation which produces an annual allowance that does not, during the first two-thirds of the life of the asset, exceed the allowance under the declining-balance method may be used.

The unit-of-production method of depreciation is one which could be used by a farmer for machinery. By this method the depreciation is computed on the basis of the amount of production attributable to the asset during the year; for example, depreciation on a combine might be measured at $x per acre used.

Some equipment may be suited to depreciation on a machine-hour or day basis. This is commonly used with construction equipment. The life of the machine is measured in hours or days of use and the depreciation is based upon that measure.

Real Estate Depreciation

Since 1969, depreciation on real estate has been limited. The only accelerated method which may be used on new buildings and other depreciable real estate is the 150% declining-balance method (or any method which does not depreciate any faster than that). Real estate includes any property described in the recapture of real estate depreciation rules. See § 5.11. Residential rental property units are excepted if there is more than one dwelling in the building.

Used real estate, buildings, structural components and the like may be depreciated only on the straight-line basis if acquired after July 24, 1969. Again, there is an exception for rental residential housing.

If a fast method of depreciation is being used on real estate, the method may be continued or the farmer may change to straight line if he desires to avoid the more rigid recapture rules on real estate. See § 5.11.

Farm Income Tax Manual, § 505.

§ 5.6 USEFUL LIFE

Rate of depreciation and the applicability of 20% first-year depreciation and the fast methods of depreciation are determined by the estimated useful life of the asset.

The Internal Revenue Service has established an elective system for determining useful lives of assets acquired before January 1, 1971 (Guideline Class System) and for assets acquired after December 31, 1970 (Asset Depreciation Range). A taxpayer need not use this system, and if he does not, he must prove the accuracy of the lives used.

Useful life is closely related to salvage value (§ 5.7) in that both are determined by reference to the individual farmer's normal period of use of an asset, not the intrinsic life of the asset nor the junk value of the asset at the end of that life. For example, if a farmer regularly uses a tractor for four years, the useful life to him is four years even though such a tractor has an inherent useful life of ten or twelve years. To him, the salvage value of the tractor is its estimated value at the end of four years. The taxpayer here must run the risk that he can successfully show that his estimate of useful life is accurate.

With the loss of the faster methods of depreciation on real estate, useful life takes on a particular importance. The shorter the useful life that can be justified, the higher will be the depreciation actually claimed each year on new or used real estate under the 150% declining-balance method or under the straight-line method. In many cases the useful life allowed under the Asset Depreciation Range system will be shorter than the facts justify, and the taxpayer should consider using ADR.

The ADR System

The ADR System applies to acquisitions after December 31, 1970. These are the rules:

1. There are Asset Depreciation Ranges. Assets are grouped in broad classes, and for each guideline class, the taxpayer may choose a useful life which is 20% below or 20% above the guideline life specified for each broad class of assets. See below. Fractional years are rounded. If there is no guideline class for an asset, ADR is not available. However, a taxpayer may apply to have a guideline class life assigned to assets. Because, trees, vines, and nearly all land improvements are not assigned a guideline life, they do not qualify for ADR unless the taxpayer applies to the Internal Revenue Service and receives a guideline class life for such assets.

2. The taxpayer may use the "half-year convention" which deems all acquisitions to have occurred on the first day of the second half of the year. There is also a "modified half-year convention" under which all purchases in the first half of the year are treated as being placed in service on the first of the year. All purchases in the second half of the year are treated as being placed in service on the first day of the next year.

3. Adjustments to salvage value are made only where the estimate by the taxpayer is not reasonable. The taxpayer's estimate will be treated as reasonable unless there are facts and circumstances existing at the end of the year in which the estimate is established which require a determination that the estimate is inaccurate by more than 10% of the unadjusted basis of the asset. This 10% is measured after giving effect to Section 167(f) which allows salvage to be reduced by 10% of the basis of the property.

Annual Election

The taxpayer may make his election each year. If made, as a general matter, the election is effective for all additions that year. There are several exceptions permitted or required. Largely, the exceptions relate to used property and Section 1250 property. Regulations § 1.167 (a)-11 should be consulted for the specifics.

An election is effective only as to property acquired after 1970.

What Property Is Eligible

Under the regulations, the term "eligible property" means new or used property which is subject to the allowance for depreciation provided by Section 167(a) but only if—

1. An asset guideline class and period are in effect for such property for the taxable year of election; (many farm assets do not have a guideline class and period; they do not qualify. But a taxpayer may apply to the Office of Industrial Economics in the Internal Revenue Service to have a guideline life assigned).
2. The property is either Section 1245 property (tangible personal property) or Section 1250 property (real estate). See §§ 5.10, 5.11 for details of the definitions. While real property is included in ADR, there is no leeway on the life to be used. ADR generally gives a 20% leeway, higher or lower than the guideline life, but not for real estate. The repair allowance shall be applied separately to each Section 1250 property which is a building.
3. The property is first placed in service by the taxpayer after December 31, 1970.

Farm assets are listed generally at paragraph 1.0 through 1.3, below. If the farm asset is not listed there, ADR cannot be applied until a guideline class life has been assigned, upon application of the taxpayer, by the Office of Industrial Economics in the Internal Revenue Service.

ASSET GUIDELINE CLASSES AND PERIODS
ASSET DEPRECIATION RANGES
ASSET GUIDELINE REPAIR ALLOWANCE PERCENTAGES
(Rev. Proc. 72-10, 1972-1 CB 721)

Asset guideline class	Description of assets included	Asset depreciation range (in years) Lower limit	Asset guideline period	Upper limit	Annual asset guideline repair allowance percentage
00.0	DEPRECIABLE ASSETS USED IN ALL BUSINESS ACTIVITIES, EXCEPT AS NOTED:				
00.1	Office furniture, fixtures, machines, and equipment: Includes furniture and fixtures which are not a structural component of the building, and machines and equipment used in the preparation of papers or data. Includes such assets as desks, files, safes, typewriters, accounting, calculating and data processing machines, communications, duplicating and copying equipment.	8	10	12	7.5
00.2	Transportation equipment:				
00.21	Aircraft (airframes and engines) except aircraft of air transport companies	5	6	7	14.0
00.22	Automobiles, taxis	2.5	3	3.5	16.5
00.23	Buses	7	9	11.0	11.5
00.24	General purpose trucks, including concrete ready-mix trucks and ore trucks for use over-the-road:				
00.241	Light (actual unloaded weight less than 13,000 pounds)	3	4	5	16.5
00.242	Heavy (actual unloaded weight 13,000 pounds or more)	5	6	7	10.0
C0.25	Railroad cars and locomotives, except those owned by railroad transportation companies	12	15	18	8.0

Asset guide-line class	Description of assets included	Asset depreciation range (in years)			Annual asset guide-line repair allow-ance percent-age
		Lower limit	Asset guide-line period	Upper limit	
00.26	Tractor units used over-the-road	3	4	5	16.5
00.27	Trailers and trailer-mounted containers	5	6	7	10.0
00.28	Vessels, barges, tugs and similar water transportation equipment, except those used in marine contract construction	14.5	18	21.5	6.0
00.3	Land Improvements: Improvements directly to or added to land that are more often than not directly related to one or another of the specific classes of economic activity specified below. Includes only those depreciable land improvements which have a limited period of use in the trade or business, the length of which can be reasonably estimated for the particular improvement. That is, general grading of land such as in the case of cemeteries, golf courses and general site grading and leveling costs not directly related to buildings or other structural improvements to be added, are not depreciable or included in this class but such costs are added to the cost basis of the land. Includes paved surfaces such as sidewalks and roads, canals, waterways, drainage facilities and sewers; wharves and docks; bridges; all fences except those included in specific classes described below (i.e., farm and railroad fences); landscaping, shrubbery and similar improvements; radio and television transmitting towers, and other inherently permanent physical structures added to land except buildings and their structural components. Excludes land improvements of electric, gas, steam and water utilities; telephone and telegraph companies; and pipeline, water and rail carriers which are assets covered by asset guideline classes specific to their respective classes of economic activity				

Asset guide-line class	Description of assets included	Asset depreciation range (in years) Lower limit	Asset guide-line period	Upper limit	Annual asset guide-line repair allow-ance percent-age
01.0 to 79.0	DEPRECIABLE ASSETS USED IN THE FOLLOWING ACTIVITIES		20		
01.0	Agriculture: Includes only such assets as are identified below and that are used in the production of crops or plants, vines and trees (including forestry); the keeping, grazing, or feeding of livestock for animal products (including serums), for animal increase, or value increase; the operation of dry lot or farm dairies, nurseries, greenhouses, sod farms, mushroom cellars, cranberry bogs, apiaries, and fur farms; the production of bulb, flower, and vegetable seed crops; and the performance of agricultural, animal husbandry and horticultural services.				
01.1	Machinery and equipment, including grain bins and fences but no other land improvements	8	10	12	11.0
01.2	Animals:				
01.21	Cattle, breeding or dairy	5.5	7	8.5	....
01.22	Horses, breeding or work	8	10	12	....
01.23	Hogs, breeding	2.5	3	3.5	....
01.24	Sheep and goats, breeding	4	5	6	....
01.3	Farm buildings	20	25	30.0	5.0

Form 4832 must be used to submit information and make the election.

Guideline Class System—Pre-1971 Assets

There is now a Guideline Class System for assets acquired before 1971. It is offered as the ADR equivalent for assets acquired before ADR became effective in 1971. This system differs from ADR by requiring that the depreciation rate be the guideline life without leeway to choose any figure 20% greater or lesser than the guideline as in ADR. Again this is an election open to the taxpayer.

Depreciation for such assets having a guideline class will be determined as follows:

(a) The property must be accounted for in one or more depreciation accounts which conform to the asset guideline class, and the depreciation for each such account must be determined by using a rate based upon a life not less than the class life, or

(b) The property must be accounted for in one or more depreciation accounts (whether or not conforming to the asset guideline class) for which depreciation is determined at a rate based upon an estimate of the lives of the assets (instead of the class life) and the total amount of depreciation so determined for the asset guideline class may not be more than would be permitted under paragraph (a) above using the method of depreciation adopted by the taxpayer for the property.

If the allowance claimed for a guideline class is in excess of a reasonable allowance, it will be adjusted as follows:

(a) If the excess is established to the satisfaction of the Commissioner to be the result of a good faith mistake by the taxpayer in determining the maximum amount permitted as described above only such excess will be disallowed, and

(b) In all other cases, the taxpayer's election to apply these rules to the asset guideline class for the taxable year will be invalid and the reasonable allowance for depreciation will be determined without regard to them.

The election is made by filing it with the tax return. An election may be made on an amended return only if filed within time for filing (including extensions). Use Form 5006.

Those are the options open under the present regulations. Each taxpayer should study them carefully because useful life is a factor in many determinations. Obviously, it determines the rate of depreciation on straight-line, declining-balance and sum-of-the-digits methods. But, in addition, it is determinative of the use of other rules. For example, fast depreciation is allowed on property having a useful life of three years or more and 20% first-year depreciation on assets with a six-year life. It is also determinative of the amount of investment credit available for an asset. See § 5.9.

Rapid Depreciation

If the tax rates and taxable income are expected to be about the same each year, the taxpayer should choose a method of depreciation which gives the highest possible deduction each year. Taxes saved this year are worth more than the same amount of tax savings in the future. Lower taxes this year mean less borrowing to pay taxes or more income to be saved. While ADR adds some complexities, it generally will yield the most early depreciation when combined with either the double-declining balance or the sum of years' digits method.

Unless there is some compelling contrary reason, the greater depreciation allowable seems worth the added detail.

§ 5.7 SALVAGE VALUE

Salvage value of an asset at the end of its life must be recognized (except where ADR is chosen) in the depreciation computation either as an adjustment to the basis of the asset or in the rate of depreciation. ADR also specifies liberal rules in adjusting salvage value. No asset may be depreciated below salvage value.

Salvage value is the amount realizable from the disposition of the property after it has become no longer useful in the taxpayer's business and is retired from use. The objective of depreciation calculations taking salvage value into account is to make the aggregate of the depreciation deductions plus the salvage value at the end of the useful life equal the original cost or other basis of the property.

If ADR is not elected, salvage value, except where a declining-balance method of depreciation is used, reduces depreciation in one of two ways: (1) by reduction of the basis on which depreciation is computed, or (2) by reduction of the rate of depreciation. In either instance the amount of net salvage should actually, or in effect, be a credit to the depreciation reserve.

Even where a declining-balance method is used or when ADR is chosen and salvage is ignored in determining the annual depreciation allowance, an asset may not be depreciated below its reasonable salvage value.

Because useful life is the period for which the farmer customarily uses an asset, regardless of the intrinsic life of the particular piece of property, salvage value is the expected sale price upon the termination of the farmer's period of use and is not related to junk or scrap value unless he customarily uses such assets until they wear out.

Salvage value is an easy issue for tax auditors to raise. If a farmer has not recognized salvage, he is placed on the defensive, since it is hard to argue in most cases that an asset has no salvage value at all.

Failure to recognize salvage value on assets currently added to the depreciation schedule is a calculated risk. Some farm-tax advisers continue to ignore salvage value. The safest approach is to recognize a minimum salvage value on all assets, unless it is certain that a particular asset will be totally consumed or is subject to the 10% rule discussed next.

If the farmer has elected ADR, salvage value may be adjusted by the Internal Revenue Service only if it is in error by 10% or more of the adjusted cost of the asset.

Examples

A farmer buys an asset for $80,000. He elects ADR. He determines that salvage value will be $16,000. He may under Section 167(f) of the Code reduce this figure by 10% of his cost to $8000. So long as actual salvage value does not exceed $24,000, the taxpayer's value will be approved. If he had not elected ADR, his salvage value could be adjusted if it exceeded $16,000, his original determination.

10% Salvage Slack

In estimating salvage value, the farmer has some leeway on assets with a life of three years or more. He may reduce the salvage value by 10% of the basis of the asset. Thus, if a tractor cost $6000, had a useful life of eight years and a salvage value of $1000, the farmer could reduce the salvage value by $600 (10% of the basis of $6000) and use $400 as salvage value in computing his depreciation.

As a practical matter, if salvage value would be 10% or less of the cost of the asset, salvage may be ignored.

The 10% rule above does not apply to livestock; the "true" salvage value must be used. Salvage value on livestock is an area of controversy. Here ADR would allow salvage to be ignored. Many farmers have used no salvage value in the past. Arguments regularly arise on whether salvage value of breeding and dairy animals should be a cutter-and-canner price or commercial-grade price or the value of the cattle for breeding or dairy purposes.

One case set the salvage value of range cows at $75 and another at fifteen cents per pound for 750-pound weights. Cutter-and-canner price is high for range cattle unless it is reduced by possible death losses and transportation costs.

In one case, a rancher whose cows had a basis of $73 each was faced with the Commissioner's contention that the animals had a salvage value of $100. In an important taxpayer victory, the court held that the variables and risks involved justified a salvage value of $45 on range cows. It was impressed with the violent cyclical and seasonal price fluctuations, the possibility of disease, a range mortality rate of about 4%, and marketing expenses of $15 or $20.

§ 5.8 BASIS FOR DEPRECIATION

Basis for depreciation is the cost to the purchaser, and the donor's basis to the donee. Basis of property received from a decedent

dying after 1976 is determined under the complex new rules of the 1976 Act. Before 1977, such property took a basis equal to the fair market value at death. Several other basis situations warrant discussion.

The basis for depreciation is the same as the basis for determining gain on the sale or disposition of property. The major practical exception to this rule involves property converted to business use.

"Basis" is a word of art in tax law meaning the amount that the taxpayer has invested in property for tax purposes—or the capital amount that he may recover free of tax. "Unadjusted basis" refers to the initial amount of his tax investment. "Adjusted basis" means the remaining amount as of any given date that the taxpayer is entitled to recover.

The unadjusted basis is adjusted upward for additions, betterments and improvements, that is, additional investment, and for disallowed deductions on capital gain transactions such as the expenses of growing unharvested crops sold with the land. It is adjusted downward for any recovery of the capital interest, for example, by deductions for depreciation or casualty losses and by the amount of severance damages received in condemnations. Basis of depreciable property must be adjusted for depreciation allowable even though it was never deducted.

Basis of Purchased Property

Basis of property purchased by a farmer is cost, including the amount of any mortgages and other obligations on the property.

"Cost" usually means the total purchase price of the asset, the cash and the value of any other property paid. For example, if a farmer bought a farm for $100,000 paying $30,000 down and giving or assuming a mortgage for $70,000, the cost basis is $100,000.

The mortgage obligation on the property is included in basis even though the purchaser is not liable on the mortgage. This underlies some tax shelter plans and invites IRS challenges. To use an oversimplified illustration, assume that a building could be acquired for $1000 cash and a no-liability mortgage of $99,000. The basis is $100,000. If the depreciation on the building is $3000, the "owner" has a tax deduction in the first year which exceeds his out-of-pocket investment. This troubles the IRS. Remember, however, that deductions from a farming activity may not exceed the amount "at risk." See § 4.1. For other activities, either the "at risk" rules or the partnership limitations on losses may apply. See § 8.3.

Where depreciable and nondepreciable properties are purchased or acquired as a unit (as in the example above, land with buildings on it), the cost or other basis of the depreciable property must be carved out of the total amount. The amount of the basis for depreciable property cannot exceed an amount which bears the same proportion to the lump sum or total cost as the value of the depreciable property bears to the value of the entire property.

In the example above, if the buildings are worth $40,000, no more than that can be allocated to buildings. A further allocation of the $40,000 must be made to give each individual building a basis for depreciation.

The Service seems to agree that in the case of new property where the taxpayer is the first user, the actual costs of the components of a building may be used by the original owner as the basis for depreciating the separate units. The same rule holds for a used building when acquired for a lump-sum consideration. The allocation must be made to components on the basis of their condition when purchased, and ADR may not be used.

On the purchase of a farm, it is necessary to carve out components such as fences, tiling, roadways, water systems, etc., for depreciation and investment credit.

Basis of Gift Property

If property is acquired by gift, the basis for depreciation and gain to the donee is the same as it would have been in the hands of the donor or the last preceding owner who did not acquire it by gift, adjusted to the date of the gift. There are several quirks to be noted. If the value of the property at the time of the gift is less than the donor's basis, then for purposes of measuring loss, the basis is the fair market value at the date of the gift, not the donor's basis. This rule produces a "no man's land" where there is neither gain nor loss if the property is sold for an amount less than the donor's basis but more than the fair market value at the time of the gift. Although it is possible for a donor to give away appreciated property and transfer to the donee the potential gain on sale, a loss cannot be so transferred.

> **Example:** A farmer gives his son a piece of land which cost $5000 and is worth $10,000. If the son sells it for $10,000, he reports the $5000 gain. If the land cost $10,000 but it was worth $5000 when it was given to the son who sold it for $4000, the loss to the son would be only $1000.

If property is received by gift and a gift tax is paid, the basis of the property to the donee for depreciation or gain is increased by some or all of the amount of the gift tax paid. Before 1977, all of the gift tax could have added to basis. After 1976, only the gift tax on appreciation in value is so added. In either case, the increase is permitted, however, only up to the fair market value; it cannot raise basis above value.

The holding period for capital gains includes the time that the donor held the property if the donee takes the donor's basis.

Basis of Property Received at Death

When the decedent died before 1977, property acquired from him took a basis equal to the fair market value of the property at death. The estate and gift tax changes made by the 1976 Act also changed this rule. Where death occurs after 1976, the takers of the property will use a "carry-over" basis for the property. "Carry over" means the same basis as the decedent had immediately before his death. This "carry over" may be adjusted by two factors. These adjustments are (1) increases in value which occurred before January 1, 1977, and (2) federal and state death taxes on appreciation in value after January 1, 1977. The adjustment for increases in value before 1977 is made for the purpose of determining gain and depreciation, but not for loss. While one of the Committee Reports to the 1976 Act said this adjustment could not increase the basis above the value at the date of death, the statute does not contain this limitation, and the statute apparently will apply. In some cases, then, this adjustment could cause basis to exceed death tax values. Some observers believe that the adjustment to values at December 31, 1976, which may result in a basis higher than death tax values, resulted from mere oversight and will be reversed by Congress in the near future.

The adjustment for death taxes is made for the purpose of determining both gain or loss. However, this adjustment may not raise the basis to the point where it exceeds the death tax value.

There are special rules for determining the increase in value which occurred prior to 1977. For marketable bonds or securities, the value at December 31, 1976, is determined by the market quotations on that date. For all other property, any increase in value is allocated prorata (1) to the period from date of acquisition to 1977 and (2) to the period from December 31, 1976, to the date of death.

Household goods up to $10,000 in value may, at the executor's election, be excluded from the carry-over basis rules and these will take a basis equal to the fair market value at death. Also, an estate having an aggregate carry-over basis of $60,000 or less (after the adjustments for 1976 values and for death taxes) may increase the carry-over basis to the lesser of fair market value or $60,000.

Some examples may help to clear up these rules. In all cases, death occurs after 1976.

> **Example 1:** The decedent paid $200 for a marketable security. Its value at December 31, 1976 was $900 and at death was $1000. Its basis is $900 plus state and federal death taxes on $100, the value at death less the value at December 31, 1976. If it had

a value of $400 at death, the basis would be $900. The adjustment for death taxes could not be made in this case.

Example 2: On December 31, 1975, the taxpayer purchased a painting for $5000. The taxpayer died on December 31, 1977. At his death, the painting had a value of $10,000. The total increase in value is $10,000 (the value at death) reduced by $5000 (the cost) or $5000. This $5000 increase in value must be divided between the time before and the time after January 1, 1977. Using the formula above, there were 366 days (1976 was a leap year) in the period before January 1, 1977. The decedent's holding period was 731 days. (The date of acquisition is not counted but the day of death is.) The formula for allocating the $5000 appreciation is:

$$\frac{366 \text{ (days before January 1, 1977)}}{731 \text{ (total days of ownership)}} \times \$5{,}000 = \$2{,}503$$

The formula value at December 31, 1976 is $5000 (cost) plus $2503 (increase in value before January 1, 1977) or $7503. This amount may be further adjusted for Federal and state death taxes attributable to the $2497 increase in value after 1976. If we assume those taxes are $997, the total basis is $7503 plus $997 or $8500.

Example 3: Assume that the asset was a farm purchased for $50,000 but which had a value of $100,000 at death. However, under the special valuation rules for farmland (§ 9.2(b)) the farm's value is only $70,000. The appreciation which may be added is determined under the formula set out in Example 1, but only $20,000 of appreciation may be allocated because the special estate tax valuation rule was used for the farm.

The formula for allocating increases in value to pre-1977 and post-1976 periods must be used for all property other than marketable bonds or securities even though the true value on December 31, 1976, can be proved in other ways.

There will be regulations to cover cases where capital improvements have been made, and they generally will be treated as separate properties.

Personal and Household Effects

The executor may exclude $10,000 of household goods or personal effects from the carry-over basis and elect a basis for them equal to their fair market value at death.

In Example 2, because the painting is a personal effect, the executor might have chosen to withhold the painting from the computation shown there and to value it at $10,000 with a basis equal to $10,000.

Carry-over Basis of Less than $60,000

If the decedent's basis plus the adjustment for the increase in value to December 31, 1976, plus the death tax adjustment attributable to the increase in value after December 31, 1976, is less than $60,000, the basis to the estate or heirs will be the lesser of $60,000 or the fair market value at death.

Example 4: The decedent dies on December 31, 1977. The death value of all property is $90,000, but the aggregate bases of property after adjusting them for increases in value before January 1, 1977, and death taxes on increases in value occurring after 1976 is $40,000. This amount is divided between $10,000 for the painting in Example 1, $25,000 for Property 1, and $5000 for Property 2. Property 1 has a value of $50,000 and Property 2 has a value of $30,000. Since the executor has chosen to value the painting at its death value, it is not entitled to any further adjustment. Basis of other properties is raised to $60,000 by multiplying the difference between $60,000 and the bases of Property 1 and Property 2 ($30,000) by a fraction. For each property, the numerator of that fraction is the amount by which the value at death of the particular property exceeds the basis of that property, and the denominator is the amount by which the death values of all carry-over basis properties exceed the bases for all carry-over basis properties. However, no personal or household goods may be adjusted to more than fair market value. In this case then, the painting is excluded from the computation because the executor has elected to value it at $10,000 and it is not carry-over basis property. For the other properties the further adjustment is:

Property 1

Death value is $50,000 and the carry-over basis as adjusted for increases in value before 1977 and death taxes on increases in value after 1976 is $25,000. The excess of the death value over carry-over basis is $25,000. All carry-over basis properties have a death value of $80,000. Only Property 1 and Property 2 are included as carry-over basis properties because the painting was excluded from this category by the executor's election. Total basis for carry-over basis properties is $30,000. The excess of value for all carry-over basis properties over the carry-over bases is $50,000. The formula then is:

$$\frac{\$25{,}000 \text{ (excess of value of Property 1 over carry-over basis)}}{\$50{,}000 \text{ (excess of value of all carry-over basis properties over all carry-over bases)}} \times \$30{,}000$$

The $30,000 is $60,000 less the total bases of carry-over basis properties (Property 1, $25,000; Property 2, $5000).

Property 2

The same computation yields a fraction as follows:

$$\frac{\$25{,}000}{\$50{,}000} \text{ x } \$30{,}000 = \$15{,}000$$

In each case, the adjustment of $15,000 is added to the carry-over basis. The painting will have the elected basis of $10,000, Property 1 will have a basis of $40,000, and Property 2 will have a basis of $20,000. All of this totals $70,000. There is a basis of $10,000 for the personal effects excluded from carry-over basis plus a minimum basis of $60,000 for carry-over basis properties because the carry-over bases were less than $60,000 but the value was more than $60,000.

Income in Respect of a Decedent

The new carry-over basis rules do not apply to items of income in respect of a decedent because these items always had a carry-over basis. The pre-1977 rules continue except that the deduction for Federal estate taxes paid on an item of income in respect of a decedent was expanded to include state estate taxes. When the decedent's income is reported by the estate, it may deduct the death taxes paid on the right to income.

List of Exceptions

The following are not carry-over basis property, and the rules relating to carry-over basis do not apply.

(1) Items of income in respect of a decedent;

(2) Life insurance proceeds;

(3) Joint and survivor annuities under Section 72 and deferred compensation benefits taxable to the beneficiary;

(4) Property included in the gross estate by Sections 2035, 2038, or 2041, which has been disposed of before the decedent's death in a transaction in which gain or loss was recognized; and

(5) The income element of stock options.

Basis of Property Received in Trade

If property is traded for other property and no gain or loss is recognized, the basis of the asset received is the same as the old plus any cash boot paid. For example, if a tractor with an original basis of $4000 depreciated down to $2000 is traded for a new tractor and the farmer pays an additional $1500 in cash, the basis of the new tractor is $3500. If a farm worth $50,000 with a basis of $25,000 is exchanged for an apartment building worth $50,000 and no gain is recognized, the basis of the apartment is $25,000. Where gain is recognized, the gain is added to basis. These transactions become a little more complicated when mortgages are assumed or the farmer receives money on the trade, but the theory that the basis of the new property is determined from the basis of the old still follows. See § 3.27.

If an asset used partly for business and partly for personal purposes is traded for a new one, the business portion of the new basis must be determined.

Basis of Replacement Property

If property is acquired as replacement property after an involuntary conversion and gain is not recognized, the basis is the cost of the replacement property less the amount of the gain not recognized. See § 3.24.

Basis of Personal Asset Converted to Business

An item used solely for personal purposes may be later converted to business use, for example, an automobile, water system, refrigerator, freezer or other asset adaptable to both personal and business use. The proper basis for depreciation or for figuring loss in such case is the fair market value on the date of conversion to business use or the adjusted basis, whichever is lower. However, for determining gain the basis is the original cost or other basis adjusted for depreciation.

Other Basis Rules

Property removed from inventory takes the last inventory value as its basis for depreciation or sale. Property actually owned by a farmer on March 1, 1913, when the income tax became effective, takes as its unadjusted basis for gain the cost reduced by depreciation sustained before March 1, 1913 or the fair market value on that date, whichever is greater. For computing loss, the basis is cost less pre-March 1, 1913 depreciation.

Farm Income Tax Manual, § 511.

§ 5.9 THE INVESTMENT CREDIT: RECAPTURE OF CREDIT

The investment credit (a reduction of tax) is allowed in an amount equal to 10% of the "qualified investment" in depreciable tangible personal property and other tangible real property (except buildings) used in farming or for farm storage or research purposes which has a useful life of three years or more.

The investment credit was first enacted in 1962. It was suspended and did not apply to assets acquired or ordered during the period from October 10, 1966 through March 9, 1967, when it was restored. Then, effective April 19, 1969, it was repealed. Finally, it was revived in 1971 to apply to orders after March 31, 1971 and acquisitions after August 15, 1971. This credit differs some from the previous one, but the general outlines are the same. The rate was raised from 7% to 10% in 1975. The rate is scheduled to fall back to 7% in 1981.

Credits are subject to recapture as additional tax if an asset is disposed of prior to the end of the useful life used in computing the credit.

The effect of the investment credit is to reduce the farmer's tax by 10% of the qualified investment in most items of depreciable property (except buildings and their structural components) acquired in the farm business or for farm research or storage purposes. The credit is taken as an offset against the computed tax in the year in which the asset was placed in service, generally the year of acquisition.

The credit is not an election; it is mandatory. It must be claimed on all eligible properties. If there is no current tax to offset, the credit may be carried back to reduce the tax for prior years and forward to reduce the tax in future years.

The measure for determining the amount of the credit is the "qualified investment," a percentage of the cost of the property determined according to the estimated useful life of the property:

If the useful life is:	The % is:
3 years or more, but less than 5	33 1/3%
5 years or more, but less than 7	66 2/3%
7 years or more	100%

The previous credit had required an additional year for each category.

Eligible Property

Generally, property eligible (defined as Section 38 property) for the investment credit is all depreciable tangible personal property (including livestock except horses) with a useful life of three or more years. Also included is other depreciable tangible property (not a building or its structural components) if (a) used as an integral part of manufacturing or production (including farming) or (b) if it constitutes a research or storage facility used in connection with these activities.

Machinery, equipment and livestock are the principal types of investment eligible for the credit but also included are: (1) tangible real estate such as tile drains, feeding floors or installations, corrals, water systems outside of a structure, outside power and light systems, telephone poles, depreciable parts of dams or drainage ways, fences, gates, paved barnyards, water wells, etc.; (2) items attached to buildings which are not part of the basic structure, e.g. fans, motors, ventilators, heaters and coolers used for environmental control; (3) storage facilities for fungible goods such as corn cribs, grain bins, silos, etc.; (4) orchards, groves and vineyards; and (5) research facilities.

Buildings and their components are not eligible. The law limits qualifying storage facilities to bulk storage of fungible commodities. Unless the work area is small in proportion to storage area, the structure will not qualify. Apparently, the restriction will eliminate combination grain and machine storage structures. Corn cribs, grain bins and silos qualify.

The word "buildings" does not include structures housing property used as an integral part of a manufacturing or production activity if the use of the structure is so closely related to the use of the equipment it houses that the structure clearly can be expected to be replaced when the property it houses is replaced. An example of a structure which is not considered a building is a confined hog system. Thus it would be eligible for an investment credit. The concept should extend to other integrated systems.

Other depreciable tangible property (not buildings) would include those items eligible under prior law, e.g., tile drains, feeding floors, water systems outside of a building, depreciable drainage ways, dams, fences, gates, etc.

While generally the use of the property must commence with the taxpayer (i.e., the property must be new), a limited amount of used property will qualify. For 1975 through 1980, a credit may be claimed for purchases up to $100,000 of used property each year. After 1980, the used property limit is scheduled to fall to $50,000 annually.

Livestock

Since 1971, depreciable livestock (other than horses) qualify for the credit. Additionally, a special rule applies to livestock. There was concern that livestock would be sold and repurchase made of identical livestock to obtain the investment credit. To block this maneuver, the law provides that, if within a one-year period starting six months before the date of acquisition, substantially identical livestock are disposed of without any investment credit recapture (livestock on which a credit was not permitted, e.g., raised livestock), the credit will be allowed only on the excess of the cost of the acquired livestock over the amount realized on the disposition. This rule does not apply if the replacement is due to an involuntary conversion—including drought and disease.

> **Example:** A taxpayer disposes of a portion of a breeding herd for $50,000 and acquires substantially identical breeding cattle within six months for $85,000. The cost of the acquired cattle qualifying for the credit is $35,000.

The definition of identical is not entirely clear. A cow culled from the breeding or dairy herd and replaced with a younger animal for breeding or dairy purposes seemingly would not be considered identical. The credit should be allowed without the limitation. If animals sold from the herd and the acquired animals are nearly the same age, then they would be considered as identical and the limitation would apply. In any event, if the investment credit had been allowed on the acquisition of the sold animals, the limitation would not apply. There could be recapture due to early disposition.

Livestock is considered new if put to its first use by the taxpayer. For example, if an animal is used for dairy purposes by one taxpayer and sold to another for breeding, it would be new for the first taxpayer and used to the second.

Recapture on Disposition

If property for which a credit was allowable is disposed of before the end of its useful life, the tax for the year of disposition is increased by the amount of credit not fully "earned" (or the credit carry back and carry over is reduced). This is the excess of the credit originally allowed over the amount of credit that would have been allowed if the estimated useful life had been the actual useful life, that is, the period from the date of placing the asset in service to the date of disposition. The tax resulting from a recaptured investment credit may not be offset by an investment credit.

Disposition includes the retirement or abandonment of an asset, conversion of an asset to personal use, gift, or transfer upon foreclosure, but not the mere leasing of property or the transfer of a security interest. If the basis of the property is reduced, for example, by a refund of part of the purchase price, there is a disposition pro tanto. If an asset is used for business and personal use, there is a disposition pro tanto as the personal use increases. For example, if a credit was claimed on 75% of the cost of a car, the business portion, and subsequently the business use fell to 50%, there would be a disposition of 25% of the qualified investment. If the business use increases, there is no additional credit.

Death is not a disposition requiring recapture. Property in a partnership, or in a Subchapter S corporation (§ 8.10), or in an estate or trust, or in joint tenancy, or otherwise owned which passes on the death of the partner, shareholder, beneficiary, joint tenant or owner is treated as if it had been held for its full estimated useful life. Nor is there a disposition when an estate or trust transfers property to a beneficiary.

No disposition occurs if the transfer is a mere change in the form of doing business, for example, a transfer to a partnership or corporation in a tax-free exchange, if the transferor retains a substantial interest in the business or an interest equal to his share in the prior form. If such a transferee or a donee disposes of the property, he (or it) is credited with the holding period of the transferor.

Indirect Dispositions

A shareholder of a Subchapter S corporation which elected to be taxed as a partnership (§ 8.10) makes a disposition requiring recapture if the corporation disposes of the asset or if the shareholder's proportionate stock interest is reduced (for example, by sale, redemption or the issuance of new shares) by more than one-third. If a shareholder transferred 25% of his stock, there would be no recapture. But if he sold or gave away 50% of his stock, he would have disposed of 50% of his share of the property subject to the investment credit.

Disposition similarly occurs if a partner's interest in partnership profits (or in the particular asset) is reduced by more than one-third. If a beneficiary's proportionate interest in an estate or trust is reduced by more than one-third, there is a disposition pro tanto of his interest in the property which was subject to the investment credit.

If a regular corporation elects to be taxed under Subchapter S, there is a disposition of its assets, unless special shareholder consents are executed.

Replacement of Property

If the taxpayer sells and replaces property, the sale is subject to recapture for any unearned credit. The purchase is, however, eligible for the credit. The same rule applies to property lost by casualty or theft. The event is a disposition subject to recapture, but the replacement qualifies for the credit. The credit in effect before April 18, 1969, had different rules for replacements.

Carry Overs and Carry Backs

The credit may not exceed the tax in any year. The amount of credit which may be used in any one year is $25,000 plus one-half of the tax in excess of $25,000. If there are excess credits in any year, they may be carried back to three prior years and over to future years subject to this limitation on amount. Before 1976, current credits were first used, then carry overs from prior years, then carry backs from later years. Beginning in 1976, the order is carry overs from prior years (first year, first), current credits, carry backs from later years. Thus, a "first-in, first-out" rule is now applied to investment credits. Carry-over credits arising before the 1969 repeal were subject to special rules governing their use in later years.

Addition to Tax

The unearned credit which is recaptured upon disposition must be treated as tax due even though, otherwise, there would be no tax liability because of a loss for the year, or a net operating loss carry over or because no return was required. The tax resulting from recapture is not reduced by the investment credit, the retirement income credit, or the credit for the elderly.

§ 5.10 RECAPTURE OF DEPRECIATION ON PERSONAL PROPERTY

On the sale or other disposition of depreciable property (except buildings and their structural components), the amount of the gain which represents the recovery of depreciation deductions taken in 1962 and thereafter will be treated as ordinary income.

Depreciation on livestock claimed in 1970 and thereafter will be subject to recapture as ordinary income up to the amount of the gain on the sale or other disposition of the animals.

Amounts which would otherwise be capital gains are converted into ordinary income on the sale of depreciable property to the extent that the gain recognized represents the recovery of depreciation deducted in 1962 or later. Despite the rules of Section 1231 or other sections, a depreciation deduction taken after 1961 cannot serve to convert ordinary income into capital gain. However, any gain in excess of the amount attributable to a recovery of depreciation is still treated as capital gain or aggregated with Section 1231 items.

The term "Section 1245 property" describes the type of property which will receive the ordinary income treatment. It means tangible or intangible personal property which is or has been subject to depreciation. There is no useful life test. Livestock was excluded before 1970. Depreciation on old or new livestock which is claimed after 1969 will be subject to recapture from the gain on the sale or disposition of livestock.

In addition to depreciable personal property, the term also includes other tangible property (not including a building or its structural components) which was depreciable while used in the farming business or as a research or storage facility in connection with the farm business. "Building" is defined in its usual sense as a structure enclosing a space within its walls, usually covered by a roof and designed to provide housing, shelter, or working or office space, etc., for example, a tenant house, garage, machine shed, warehouse, or barn.

Thus, any depreciable item used in the farm business which is not properly a building is subject to the ordinary income treatment. Fences, water installations, outside light and power systems, depreciable parts of dams and ponds, pavements, etc., are tangible and depreciable but are not buildings or components. While a storage facility could have the characteristic of a building, as a warehouse does, most farm storage facilities, including corn cribs, grain bins, silos, etc., are not buildings and are Section 1245 property.

How to Compute Recapture

The amount to be treated as ordinary income is the lesser of the gain realized or the depreciation taken after 1961. If an asset was acquired in 1962 or later, the amount of recapture will be the depreciation taken or the gain, whichever is the lower. If the gain is greater than the amount of the depreciation recaptured, the excess gain will be treated as a capital gain or a Section 1231 gain, as the case may be.

With respect to livestock, the amounts to be treated as ordinary income are the depreciation deductions claimed for periods after December 31, 1969 but not in excess of the gain realized on the animal. On existing livestock owned as of the beginning of 1970, there will be no recapture of depreciation taken before 1970. If livestock is acquired after 1969, all of the subsequent depreciation is subject to the recapture rules.

Although the recapture provision will apply primarily to purchased livestock because raised animals in the hands of a cash-basis taxpayer have no basis, recapture will apply to the extent any livestock has a basis and is depreciated. For example, a cash-basis farmer may have inherited animals raised by his predecessor, or an accrual-basis taxpayer may have included the animal in inventory and later removed it to a depreciation schedule or he may have capitalized the costs of raising animals to avoid the "excess deductions account" recovery of farm losses. See § 5.13.

If the transaction which brings about recapture is one in which neither cash nor other property is received, the gain is measured by the difference between the basis of the property and its fair market value. On the sale of a partnership interest or a liquidation of the interest, the partner's share of depreciation recapture on assets still held by the partnership must be treated as ordinary income to him.

Generally, the depreciation deduction actually claimed will be used, but if the farmer cannot prove the amount taken, he will be charged with the amount he would have been allowed.

If property is sold on the installment method (§ 3.26), the gain in each installment is treated as ordinary income until all of the depreciation is recaptured; thereafter, any capital or Section 1231 gain is reported.

See § 5.13 on the relationship to EDA recapture.

Death or Gift

There is no recapture income recognized by the transferor on an inter vivos gift. But the donee will report all of the ordinary income when he disposes of the property. There is no income recognized to the decedent on a transfer at death. Before 1977, the estate, heir, legatee or other recipient received a new basis on death so there was no carry over of the gain that would have been reported by the decedent, except in those situations in which the decedent had sold the property before death and the proceeds are reported as income by the estate. After 1976, if there is a carry-over basis, the successor to the decedent will recapture the decedent's depreciation taken after 1976. See § 5.8.

If property is donated to a charitable organization, the amount of the deduction is reduced by the amount which would have been treated as ordinary income under Section 1245 if the property had been sold for its fair market value.

Other Dispositions

If the Section 1245 property is disposed of in a tax-free exchange or by involuntary conversion, the amount of ordinary income to be reported may not exceed the gain actually recognized on the transfer plus the value of property received which is not depreciable property and which is not otherwise included in the recognized gain. The remainder of the recapturable depreciation is carried over into the new property to be recognized when it is sold.

If the property is transferred between partner and partnership or corporation and shareholder in a tax-free transaction, ordinary income is recognized by the transferor only to the extent of the recognized gain, if any. The rest of the ordinary income representing recapturable depreciation will be recognized by the transferee if it disposes of the property at a gain.

Conversion of a business asset to personal use is not a disposition which causes recapture of depreciation. The recapture will occur when the asset is sold.

Farm Income Tax Manual, § 514.

§ 5.11 RECAPTURE OF DEPRECIATION ON REAL ESTATE

Some or all of the depreciation claimed on real estate in excess of straight-line depreciation will be recaptured as ordinary income on sale or other disposition of the property up to the amount of the gain realized.

Excess depreciation claimed from 1963 through 1969 will be recaptured according to a formula based on the period of time the property was held. If the property was held for ten years, there is no recapture.

Excess depreciation claimed after 1969 is fully recaptured without reference to the period that the property was held.

The 1962 law which provided for the conversion to ordinary income of the part of the gain on depreciable property that represented recovery of depreciation did not include all depreciable real property. See § 5.10. In 1964, Section 1250 was added to the Internal Revenue Code to apply a recapture theory to the buildings, structural components, and other real property which were excluded from the previously enacted recapture provisions.

The purpose of the real property recapture provisions is much the same, but the technique is different. The statute was aimed at the taxpayer who used the fast methods of depreciation on buildings or other real property and then sold the property for capital gains.

The original provision, as enacted in 1964, recaptured only depreciation in excess of straight-line and only a part of that if the property was held for twenty months or more. There was no recapture at all if the property was held for ten years.

In 1969, these provisions were tightened. The law still limits recapture to the excess of fast depreciation over straight-line but it eliminates the formula which recaptured a declining amount of the excess depreciation if the farmer held the property for a period from twenty months to ten years. All of the excess depreciation claimed in 1970 and thereafter will be recaptured against the gain on the sale or other disposition of real property.

Both systems exist at the same time. Excess depreciation claimed from 1963 through 1969 is recaptured under the old rules. Excess depreciation claimed after 1969 is recaptured under the new rules. Thus, there will be two computations on any real estate owned before January 1, 1970. For example, if a building was acquired in 1964, depreciated under the double declining-balance method and sold in 1977, the excess depreciation from 1964 through 1969 would not be recaptured because the property was owned for over ten years. The excess depreciation claimed from 1970 through 1977 would be recaptured against the gain on the sale.

A farmer who uses straight-line depreciation on his buildings will not fall under this section at all, with one exception—if the property is held less than a year and there is a gain on the sale, any depreciation claimed is converted into ordinary income to the extent of the gain.

The recapture rule applies to sales, exchanges, and involuntary conversions, for example, condemnation. If the property was sold on the installment basis, the gain in each installment will be treated as ordinary income until the depreciation is recaptured; thereafter, the gain will be treated as capital gain or Section 1231 gain, as the case may be.

There are different, more liberal, recapture rules for depreciation taken on residential rental units after 1969. These were also tightened by the 1976 Act.

Additional Depreciation

The real estate provision does not recapture all of the depreciation recovered in the sale price; it reaches only the part of the gain which represents recovery of depreciation taken after 1963 in excess of straight-line depreciation. For example if a building had an adjusted basis of $50,000, was sold for $60,000 and the excess of double declining-balance depreciation over straight-line depreciation was $6000, the $6000 would be subject to the recapture rules. All of the post-1969 depreciation will be recaptured as ordinary income. The depreciation claimed before 1970 and after 1962 will be recaptured according to a declining scale based upon the period of time the property was held.

The gain on the sale or other disposition of the real estate is applied first to the post-1969 depreciation and that is recovered in full. If there is any remaining gain on the sale or other disposition, the pre-1970 excess depreciation is recaptured against the gain using the formula below. If there is still any gain remaining after the recapture of the post-1969 excess depreciation and the pre-1970 excess depreciation, the gain is a capital gain or a Section 1231 gain.

Pre-1970 Depreciation

If the property was held for twenty months or less, all of the recovery (i.e., the lesser of pre-1970 "additional depreciation" or the gain) is ordinary income. If the property was held for ten years or more, none of the excess depreciation is treated as ordinary income. If the property was held for more than twenty months and less than 120 months the amount of the recovery (i.e. the pre-1970 depreciation or the gain, whichever is the lesser) to be reported as ordinary income is reduced by one percentage point for each month over twenty which the property was held. For example, if a barn which produced the "additional depreciation" recovery had been held for five years (sixty full months), the amount of the recovery of additional depreciation to be recognized as ordinary income would be reduced by 40% (one percentage point for each month over twenty). Thus only 60% of the amount that represented the excess of pre-1970 fast depreciation over straight-line (or the remaining gain, if that is less) would be treated as ordinary income.

The holding period begins the day after the day of acquisition. If property is constructed or reconstructed, the holding period begins on the first day of the month in which the property is placed in service. It includes the period the property was held before 1964 and after 1969, and the period the property was held by a predecessor, such as a donor, who recognized no gain or loss on the transfer. But a new holding period starts on property acquired in an exchange or after an involuntary conversion.

Death or Gift

No income is recognized to the donor if the property is given away, but the donee will recapture the donor's and the donee's excess depreciation against the gain when he sells the property. Even if the donee never used the property for business and so took no depreciation himself, he will recover the donor's depreciation.

Before 1977 there was no income recognized to the decedent at death nor did the estate or the devisee assume the income potential; the successor started fresh unless the property had been sold by the decedent before his death and the income had not been reported by him. After 1976 the successor to the decedent will recapture the decedent's excess depreciation deduction attributable to periods after December 31, 1976 if the property has a carry-over basis.

Property Included

The real estate recapture provision is phrased to cover all property excluded from the Section 1245 recapture rule. Thus, it applies to all intangible real property, such as leases of land or buildings, and to all tangible real property, including buildings and their structural components and any other property which by definition is not included under Section 1245. See § 5.9. In essence, the real property recapture provision excludes tangible real property used as an integral part of farm production or used as a research or storage facility in farming. Recapture would apply to a part of a residence used for business purposes.

Where a basic property has had substantial additions or improvements, these are treated as separate elements. Special rules apply to purchased leaseholds and leasehold improvements amortized too rapidly.

Tax-free Transfers and Exchanges

Generally, tax-free transfers produce recapture income only to the extent of the gain recognized. Ordinarily, a transfer of property to a new partnership or a new corporation is a tax-free transfer, so there is no recapture income; the partnership or corporation will report the recapture income if it sells the property. On a distribution of property to a partner the potential income is carried over to the partner who recognizes it when he sells the property. But on sale of a partner's interest in the partnership, he is treated as selling a share of the partnership real estate and must report any recapture.

On tax-free exchanges, for example, a farm for an apartment, or on involuntary conversions where the property is replaced, recapture income is recognized to the extent of the gain and the rest of the potential recapture is carried over into the new structure. But if nondepreciable property is acquired, for example, an exchange of a farm with buildings for bare land, the recapture is reported.

The opportunity to avoid the immediate recognition of the ordinary depreciation recapture income and the ordinary income recapture of soil and water conservation expenditures and land-clearing expenses (§§ 5.12, 5.13) through the use of the tax-free exchange of property will be important to many farmers with potential recapture income.

Comparison of Sections 1245 and 1250

It will make a difference whether property falls under Section 1245 or Section 1250. For example, if it is determined that a farm storage structure is a building under Section 1250, then the recovery is the lower of the gain or the amount of fast depreciation in excess of straight-line depreciation taken after 1963 and the 1963-1969 excess depreciation is further reduced if the building was owned over twenty months before sale of the farm. If the building was owned for ten years, there would be no recovery of pre-1970 excess depreciation. On the other hand, if the structure is a storage facility under Section 1245, the ordinary income would be the amount of the gain which represented recovery of any depreciation taken after 1961.

Note on Planning

Farmers have learned to plan the use of fast depreciation on real estate against a statute which freed them from recapture if the property was held for ten years. That planning notion must change because all excess depreciation after 1969 will be recaptured against the gain whenever the farm is sold.

If a farmer finds himself in this recapture trap, there is a partial escape. He may, without permission, change his depreciation method from double declining-balance or sum-of-the-digits to straight-line. He cannot, of course, retroactively wipe out the potential recapture of prior years, but he can rely on time to decrease or eliminate the 1963-1969 excess depreciation recapture and he can change his method for future years.

In pondering the possible recapture on the future disposition of his farm, he must be aware not only of the depreciation recapture discussed in this section and in § 5.10, but also of the potential recapture of deductions for land expenses. See §§ 5.12, 5.13.

Usually, the earlier deduction through fast depreciation will have reduced taxes, and merely because it was realized first, that earlier tax savings may have been worth more than the tax on recapture. Also the recapture income may be averaged and perhaps reduce the tax on it.

Farm Income Tax Manual, § 515.

§ 5.12 RECAPTURE OF LAND DEDUCTIONS

The Tax Reform Act of 1969 added a new recapture provision to convert to ordinary income some or all of the deductions claimed for soil and water and land-clearing expenses to the extent of the gain realized on the sale or other disposition of land.

Code Section 1252 recaptures as ordinary income on the sale or other disposition of farmland some or all of the deductions previously claimed for soil and water conservation expenditures and land-clearing expenses. The amount of ordinary income recaptured is limited to the lesser of the gain or a percentage of the deductions based upon the period of time that the land was held.

The deductions for soil and water conservation expenditures and land-clearing expenses are elective; the expenses are to be capitalized unless the farmer elects to deduct them. The election to deduct or capitalize soil and water conservation expenditures is made in the first year in which the taxpayer has any. If he made the election to deduct these expenses in the past, the election is still binding and he must continue to deduct them unless he obtains permission to change that particular aspect of his accounting. The land-clearing expense deduction election is made annually and the taxpayer is not bound from year to year.

Recapture applies to dispositions after 1969 but only with respect to deductions taken after 1969. Thus, except for the farmer who has made the soil and water conservation deduction election, a determination as to whether to deduct these expenses must be made in the light of the recapture rules.

Amount Recaptured

The amount or the portion of the aggregate deductions which will be recaptured turns on the period of time that the land was held before disposition. If the land was held for ten years, there is no recapture at all. If the land was held for five years or less, all of the deductions are recaptured up to the amount of the gain realized.

If the land was held for less than nine years and more than five, an "applicable percentage" of the deductions is recaptured. A sale or disposition within the sixth year will recapture 80% of the total deductions, within the seventh year 60%, within the eighth year 40% and within the ninth year 20%. In no case, of course, can the recapture exceed the actual gain on the disposition of the land. Note that contrary to the old rules (pre-1970) on recapture of depreciation on real estate, the "applicable percentage" applies only to the aggregate deductions and not to the lesser of the deductions or the gain.

In determining the holding period, the time of ownership before 1970 counts. Practically, a farmer who has owned his land for nine years is immune from recapture of land deductions taken after 1969.

Disposition

Disposition for purposes of this recapture provision means the same as disposition for purposes of recapture of depreciation on personal property. See § 5.10. As directed by the statute, the commissioner has issued regulations similar to those of Section 1245 for the recapture of land deductions. Following those patterns, the land recapture provision will not apply at death; the recapture potential will generally carry over to the transferee taxpayer on gift, formation of a corporation, and transfers between partners and partnership. If any gain is recognized on such transfers, the recapture will arise to the extent of the gain recognized.

It seems clear that no part of land deductions is recaptured by the decedent's estate or other successor.

On tax-free exchanges and involuntary conversions, there will be a recapture to the extent of the gain recognized and the value of nonrecapture property acquired, and the remaining recapture potential will be carried over to the new property. If there is a charitable contribution of the land, the charitable deduction will be reduced by the amount that would have been recaptured on sale.

If farmland subject to recapture is sold under the installment method, the income in each installment will be treated as recapture income until all of it is reported.

It would seem to follow that if a portion of the land was sold, the deductions should be allocated to the part sold and the part retained for purposes of recapture.

The recapture rule applies only to deductible soil and water conservation expenses and land-clearing expenses. Amounts which were not deductible because they exceeded the allowable deduction limits are not recapturable, of course. Soil and water conservation expenditures in excess of the deduction limit may be carried over and deducted in future years. A carry over from years before 1970 is not recaptured on a subsequent sale. There is no carry over for land-clearing expenses. See §§ 4.13, 4.14.

Overlap with EDA

There is an overlap between the land expense recapture provision and the new section which recaptures farm losses in the "excess deductions account." On the sale of land, the latter reaches only the lesser of the gain or the land expenses deducted in the last five years and then only if there is a balance in the EDA. Recapture of land expense deductions is first made against the EDA and any excess available for recapture falls under this section.

It should be noted that the applicability of the EDA recapture is limited because it generally applied only if farm losses in any year exceeded $25,000 and then only if the taxpayer had $50,000 of nonfarm income. Furthermore, after 1975, no additions are to be made to EDA. But the regular land expense recapture rule applies to any farmer who took a land expense deduction on land that he did not hold for nine years.

§ 5.13 RECAPTURE OF FARM LOSSES—EDA

Between 1970 and 1976 (ending with 1975), a cash-basis taxpayer had to add the excess of a farm loss over $25,000 to an "excess deductions account" (EDA) in any year in which he had nonfarm income of $50,000. The amount in the EDA was recaptured as ordinary income on the sale of farm assets which would otherwise produce capital gains. There are special rules for Subchapter S corporations.

Regular corporations and trusts maintained an EDA of all farm losses regardless of the amount of other income.

The balance in the "excess deductions account" at the end of each year will be recaptured as ordinary income against the gains

on the sale of "farm recapture property"—assets used for farming (except buildings) which ordinarily produce capital gains.

While the 1976 Act did not abolish any EDA's in existence, no additions will be made to them in any year after 1975.

The major effort of the Tax Reform Act of 1969 to knock out the tax shelter in farm investments was Section 1251 which added to the law the concept of the "excess deductions account." The so-called EDA was used to record farm losses which would be recaptured as ordinary income on the sale or other disposition of most farm assets which would be entitled to capital gain treatment under Section 1231. The provision affected very few taxpayers who owned farms.

Accrual-Basis Exemption

The provision was directed mainly at individuals who sought farm tax shelters but the statute applied to all taxpayers including Subchapter S corporations, regular corporations, and trusts. Because a taxpayer using the ordinary cash-basis rules was the one who stood to profit from the opportunity to deduct losses against ordinary income and sell farm assets at capital gains, the statute exempted a taxpayer from the operation of the EDA if he elected to keep inventories and capitalize all expenditures which were properly chargeable to capital account including those which he could ordinarily elect to deduct.

To qualify for the election a cash-basis taxpayer had to change his method of accounting to one which used inventories and capitalized expenditures. Ordinarily, this meant a change to the accrual method. See § 6.1.

Who Kept an EDA—Before 1976

An individual cash-basis taxpayer who had farm losses in excess of $25,000 in any year and also had over $50,000 of nonfarm adjusted gross income had to establish an excess deductions account. Married couples filing separate returns used $12,500 and $25,000 exemptions unless one spouse had no nonfarm adjusted gross income. If a joint return were filed, the EDA addition was the farm loss in excess of $25,000 but only if the combined adjusted gross income was over $50,000.

Similarly, a Subchapter S corporation on the cash basis which had a farm loss of over $25,000 and $50,000 of nonfarm taxable income kept an EDA. The income of the shareholder having the greatest nonfarm income was added to the Subchapter S corporation's nonfarm income. If the combined nonfarm income exceeded $50,000, the farm losses in excess of $25,000 were added to the EDA. Furthermore, if an individual shareholder of a Subchapter S corporation had a net farm loss (i.e., his farm deductions exceeded his gross income on the Schedule F) in his taxable year with which or within which the corporation's year ended, the corporation was not entitled to use the $25,000 and $50,000 limitations. It then established its EDA (or added to it) using the actual farm loss and without regard to the amount of nonfarm income. The same rule applied if any shareholder was a shareholder in another Subchapter S corporation which had a farm loss for the year. Neither the $50,000 nonfarm income nor the $25,000 farm loss limitation was applicable.

A regular corporation or trust was not entitled to the benefit of the dollar exemptions. It established an EDA (or added to it) for any farm loss without regard to the amount of nonfarm income.

No additions are to be made to any EDA after years beginning in 1975. Existing EDA's were not abolished, and they will operate to convert capital gain to ordinary income until the EDA is reduced to zero by farm income or by being used against capital gain.

What Is Included in the EDA

The Excess Deductions Account was a running record of the accumulated net farm losses of a taxpayer. A "farm net loss" meant the amount by which farm business deductions exceeded gross income from the farm business. Gains and losses on the disposition of Section 1231 assets were not included in farm gross income. An individual or a Subchapter S corporation, however, included in the EDA only the farm net loss in excess of $25,000 in any year. If a farm loss was $20,000, it was never included in the EDA and never recaptured. If the loss was $40,000, only $15,000 would have been entered in the EDA. The nonfarm income exemption also came into play. If the farm owner had $49,000 of nonfarm adjusted gross income, it did not matter what his farm loss was; it was not added to the EDA. If he had $50,001 of nonfarm income and $40,000 of farm loss, the entry in the excess deductions account for the year would have been $15,000. Nonfarm adjusted gross income was computed without regard to farm income or deductions.

If the taxpayer was a regular corporation, or a Subchapter S corporation in a year in which a shareholder had any farm loss, or a shareholder who was a shareholder in another Subchapter S corporation with a farm loss, or a trust, the dollar exemptions did not apply, so the amount to be added each year to the EDA was the actual net farm loss. Generally the net farm loss was the loss shown on Schedule F.

Continuing EDA Record

The EDA applied to the first year after 1969 in which the taxpayer had a net farm loss that was to be accounted for. In each year thereafter, the accountable net farm loss was added to the EDA. The account was decreased by the amount of farm net income in any year. If a deduction which did not result in a reduction of tax was included in the farm loss, it was not added to the EDA. The account is further reduced at the end of each year by the amount which is recaptured and reported as ordinary income for the year. The remaining balance at the end of each year is carried over to the next year.

Although "farm net loss" and "farm net income" are defined in the statute as the loss or income from the business of farming, the latter phrase is not defined. The business of farming is broadly defined to include any trade or business which qualifies for the special farm options in reporting income and expenses including the election to deduct soil and water conservation and land-clearing expenses. The statute does say that the business of farming includes the racing of horses if the taxpayer raises horses. If the taxpayer is engaged in several farm operations, they are to be treated as a single business in determining income or loss for EDA purposes.

How the Recapture Works

On the sale or other disposition of "farm recapture property" any gain will be treated as ordinary income to the extent of the balance in the excess deductions account at the end of the year. "Farm recapture property" is Section 1231 property (except land and buildings)—i.e., depreciable property held for more than six months (nine months in 1977; one year thereafter), cattle and horses held for 24 months or more and other livestock held for twelve months or more and held for draft, breeding, dairy, or sporting purposes, and unharvested crops sold with the land—if the assets are or have been used in the business of farming.

No EDA recapture ever occurs on the sale or disposition of a building or other Section 1250 property. See § 5.11. There is a special recapture rule for land; gain on the disposition of land will be recaptured to the extent of the deductions for soil and water conservation and land-clearing expenses taken in the taxable year and the four previous years, limited, of course, to the current balance in the EDA and to deductions taken after 1969. The overlap of the excess deductions account recapture and the land deductions recapture has been noted in § 5.12. Conservation and land-clearing expenses are applied first to the EDA recapture and are not again recaptured by the other provision. However, if there is no EDA or if the balance in the EDA does not recapture all of these expenses, the land deductions will be recaptured under the regular provision.

Although only losses incurred after 1969 are recaptured against gains realized after 1969, the property disposed of may be property acquired before 1970.

Amount Recaptured

The amount to be recaptured is simply the lower of the gain on the sale or disposition of any farm recapture property and the balance standing in the excess deductions account at the end of the year. There is no attempt in the statute to relate an expense or a loss to the asset with respect to which the expense or loss was claimed. The accumulated farm losses or some part of them are recovered against the gain from any farm recapture property. For example, a farm loss resulting from deductions for land improvements may be recaptured from the gain on the sale of breeding livestock. Similarly, a loss resulting from expenses of raising livestock may be recaptured from the gain on the sale of an unharvested crop. But after 1975 no losses will be added to EDA.

This recapture may overlap with Section 1245 recapture. The Service position is that Section 1245 first applies. If there is further gain, then Section 1251 (EDA) applies. The Service does not, however, give credit in the excess deductions account for Section 1245 gain reported, i.e., such gain does not reduce the balance in the EDA. It does, however, affect the amount of the net farm loss for the year, if any.

Dispositions

The dispositions that will cause a recapture of the EDA are sales, exchanges, involuntary conversions, and distributions by a corporation. There is no recapture on a transfer at death or a transfer by gift. The recapture potential and the EDA disappear on death, unless there was a pre-death sale which generated income in respect of a decedent on which the estate or successor would report the recapture income of the decedent.

The gift transfer raises a special problem of what happens to the donor's EDA. If the potential gain (the value of the property over its basis) on farm recapture property given away during any one-year period is more than 25% of the potential gain on all of the farm recapture property held by the donor before the first gift, each donee succeeds to a proportionate share of the donor's EDA. If the amount is less than 25%, the donor retains his full account. In the case of land, the potential gain is just the amount of land deductions that would be recaptured, i.e., those taken in the current and four prior taxable years.

Thus, it is possible for a donor to divest himself of some or all of his EDA recapture potential by giving property away. Presumably, he would try to give the EDA to a taxpayer in a considerably lower tax bracket.

On a tax-free exchange or an involuntary conversion where the property is replaced, recapture will occur to the extent of the gain recognized plus the value of any property which is not farm recapture property. Otherwise the recapture will occur when the acquired property is disposed of. Because the statute defines "farm recapture property" to include any property which has a carry-over basis, if a farmer traded farmland for a motel, the motel would be farm recapture property and the EDA would be recaptured on the sale of the motel.

Partnerships

A partnership does not maintain an EDA. Each partner took into account his share of partnership farm losses, gains from dispositions of farm recapture property, and other items which he aggregates with similar items of his own.

Except as otherwise stated below, recapture will occur on contribution of farm recapture property to a partnership to the extent that the fair market value of the farm recapture property exceeds the fair market value of the partner's partnership interest attributable to such property. For example, if a farmer transferred $30,000 worth of raised breeding cows with a basis of zero to a partnership (which had no other farm recapture property) in exchange for a one-third interest, he would recapture $20,000 of his EDA. In effect, he is treated as if he sold a two-thirds interest in the cows to his partners. When the partnership sells the cows he will report the remaining $10,000 as his share of the gain from the cows and will thus have accounted for the full $30,000 of potential gain. The recognition of gain on contribution of the cows can be avoided if the partnership agreement provides that the contributing partner will report all of the gain when the partnership sells the cows.

On the distribution of farm recapture property by a partnership to a partner or on the sale of a partnership interest, the potential gain on farm recapture property is treated as an unrealized receivable under the regular partnership rules. The effect is to bring about the recognition of the recapture income when the partner releases his interest in the farm recapture property. See Chapter 8.

Corporations

If a corporation is formed tax free by a transfer of farm recapture property in exchange for stock, the stock received becomes farm recapture property and the recapture will arise on the disposition of the stock. Thus, recognition of the EDA gain can be postponed by the formation of a corporation. If the farm recapture stock is given away, the gift rules noted above should apply.

Sometimes on the formation of a corporation, some gain is recognized if the stockholder receives property other than stock or securities. If this happens, the EDA recapture occurs to the extent of the gain recognized.

The corporation does not assume any part of the contributing shareholder's EDA. The potential recapture of the shareholder is transferred to his stock. The corporation as a separate taxpayer would have an EDA of its own if it suffered any net farm loss during the existence of the EDA rules.

§ 5.14 DEPLETION OF NATURAL RESOURCES

A reasonable allowance for the exhaustion of mineral deposits, timber and other natural resources is deductible as depletion. A person who has an economic interest in the resource recovers the cost of the wasting asset through the depletion allowance. It is claimed on a cost basis or as a percentage of gross income from disposal.

A natural deposit or resource is a wasting asset in that the use or sale of the resource consumes it. The sale proceeds represent both income and return of capital to the owner. The cost or investment of the owner of the resource is recovered through an allowance for depletion. The depletion allowance may be based upon the cost of the resource or it may be claimed as a percentage of the gross income from the sale of the mineral. There is no percentage depletion for timber.

Cost depletion allocates to each unit of production a prorata portion of the original cost or investment in the resource. In general, the total number of units in the deposit or timber tract (tons, barrels, board feet, cubic yards, etc.) is divided into the cost of the resource to obtain the amount of cost which should be deducted for each unit of production sold.

Percentage Depletion

Percentage depletion is applicable to all minerals. On oil and gas it applies only to so-called small producers. This rule was enacted in 1975. A flat percentage of the gross income from the property each year is treated as the depletion deduction. The percentage depletion rates are set out in the Internal Revenue Code. They range from a top of 22% of gross income from oil and gas wells to 5% for sand and gravel. Examples of rates and deposits to which they apply follow:

(1) 22%—oil and gas wells of small producers, or certain natural gas, sulphur and uranium and a long list of other deposits if located within the United States

(2) 15%—gold, silver, copper, iron ore and oil shale if from deposits within the United States

(3) 14%—metal mines outside the United States, rock asphalt, vermiculite, ball clay, bentonite, china clay, sagger clay and clay used for purposes dependent on its refractory properties

(4) 10%—coal, lignite, sodium chloride, asbestos outside the United States, and others

(5) 7 1/2%—clay and shale used for sewer pipe and brick; clay, shale or slate used as sintered or burned lightweight aggregates

(6) 5%—gravel, sand, stone, peat and others

(7) 14%—"all other minerals" not specifically named.

A "small producer" is entitled to depletion on oil and gas. A small producer is one who does not produce more than 12,000,000 cubic feet of gas or 2000 barrels of oil per day. These figures are gradually reduced to 6,000,000 cubic feet of gas and 1000 barrels of oil by 1980. The allowable percentage rate is also being reduced to 15% over a period ending in 1984.

The depletion deduction under the percentage depletion method cannot exceed 50% of the taxable income from the property computed without the deduction for depletion. Except for small producers of oil and gas, the 50% figure is raised to 65%.

The taxpayer must use the higher of cost depletion or percentage depletion. Determination of the higher is made each year.

Farm Deposits

A farmer would most likely be involved with sand and gravel, timber, sod, soil, etc. While he is entitled to take depletion on timber, he will probably use the special capital gains provisions applicable to the cutting or sale of timber. See §§ 3.17, 3.18. Farmers may report income from sand and gravel, top soil, loam and fill dirt on Schedule F and claim depletion on the expense schedule.

Sod, soil, dirt, turf, water and mosses are not included within the terms "all other minerals" for purposes of using percentage depletion. Sod and balled nursery stock are not depletable at all; the others would be subject to cost depletion if the resource is actually consumed.

Peat moss (or moss peat) is a type of peat and is subject to depletion at 5%. But no percentage depletion is allowed for moss until it becomes peat moss.

Ground water has been held to be a depletable asset where it could be shown that the rate of use exceeded the rate of recharge. The Internal Revenue Service states that it will follow the Texas case allowing depletion on ground water but will limit it to situations similar to those in the southern high plains of Texas and New Mexico where water is extracted from the Ogallala formation. The Service has issued guidelines for computing depletion by irrigators in those areas.

Two courts have authorized a depletion deduction of 27 1/2% (now 22%) of the gross income from sales of geothermal steam on the theory that it is a depletable gas.

A farmer who leases a deposit to an extractor will normally apply the proper depletion percentage against the gross royalties that he receives. As he enters the field of the operator, the depletion problems become more complex and are beyond the scope of this discussion.

Farm Income Tax Manual, § 513.

See following pages for Quick Reference charts to Rules on Claiming Depreciation and Rules on Recapture.

QUICK REFERENCE CHART TO RULES ON CLAIMING DEPRECIATION

	Property	Useful Life	Basis
	See § 507	**See § 509**	**See § 511**
Additional First-Year Depreciation (election) **See § 503**	Tangible personal property purchased, new or used. Purchases from "related" persons do not qualify.	6 years or more.	Cost. Does not apply to property acquired by death or gift or which has basis of transferor.
Ordinary Depreciation (mandatory) **See § 505**	All depreciable property, new or used.	Over 1 year.	Adjusted basis reduced by 20% additional first year depreciation and salvage.
Fast Depreciation (election) **See § 505**	Depreciable tangible property; original use must commence with taxpayer. Not available on real estate except residential rental housing.	3 years or more.	Same; except no salvage for declining-balance. But see ADR rules.
Reduction of Salvage (election) **See § 510**	Depreciable personal property. Excludes livestock.	3 years or more.	Cost or other adjusted basis.
Asset Depreciation Range ADR	§ 1245 property § 1250 property See more detailed rules at § 509 Post-1970 acquisitions new or used.		Cost reduced by 20% additional first year depreciation. Salvage value may be ignored, but no account may be depreciated below salvage value.
Class Life System	Pre-1971 acquisitions.		Effort to permit something similar to ADR for pre-1971 acquisitions. See § 509.

Method	Exchanges	Limitations	Special Rules
	See § 346		
One time deduction—up to 20% of cost.	Applies only to cash boot paid.	$4000 deduction per year on joint return; $2000 on others.	Taken only in first year of regular depreciation. Does not apply to trusts.
Straight-line, 150% declining balance.	Use basis of property acquired, usually basis of old plus boot.	Only 150% declining-balance on new buildings. Only straight-line on used buildings. Residential rental units have special rules.	Salvage value recognized except under declining-balance, but see ADR rules below. On personal property (except livestock), ignore salvage up to 10% of cost or other basis.
Double declining-balance, sum-of digits and any other not faster than double declining-balance.	Use basis of property acquired. New property only.	Not allowed on buildings. Residential rentals have special rules.	Change from double declining-balance to straight-line at will; otherwise, method may not be changed without permission. But see ADR rules below.
Reduce salvage value by 10% of cost or other basis.	Use basis of property acquired.	None.	Salvage is estimated sale price at end of taxpayer's normal period of use of asset. But see ADR rules below for more liberal rules.
Straight-line, declining-balance, sum-of-year digits, but not units of production.			May elect repair allowance described at § 402. May change declining-balance method to S-Y-D or from declining-balance or S-Y-D to straight line without consent of Commissioner. Choice of half-year or modified half-year convention.
			Salvage value not adjusted unless more than 10% in error after taking into account 10% allowed under election described at § 510.

QUICK REFERENCE CHART TO RULES ON RECAPTURE

	What Recaptured	Amount Recaptured
Recapture of Depreciation on Property other than Real Estate. IRC Sec. 1245 See § 514	Depreciation claimed on tangible and intangible personal property and other tangible property used in farming or for farm storage or research (except buildings). Includes livestock after 1969.	Lesser of gain or depreciation after 1961 (livestock after 1969).
Recapture of Depreciation on Real Estate deducted before 1970. IRC Sec. 1250 See § 515	Excess of fast depreciation over straight-line claimed on buildings and other realty not included under Sec. 1245 above.	"Applicable percentage" of lesser of gain or excess depreciation from 1964 through 1969.
Recapture of Depreciation on Real Estate deducted after 1969. IRC Sec. 1250 See § 515	Same as above.	Lesser of gain or excess depreciation claimed after 1969.
Recapture of Land Expenditures. IRC Sec. 1252 See § 517	Soil and water conservation expense deductions under Sec. 175 and land-clearing expense deductions under Sec. 182.	Lesser of gain or percentage of land expense deductions based on period land was owned.
Recapture of Farm Losses in Excess Deductions Account (EDA). IRC Sec. 1251 See § 516	Accumulated farm losses of individuals and Subchapter S corporations in excess of $25,000 in years in which nonfarm income exceeds $50,000. All farm losses of corporation, estate or trust. No additions will be made to this account in years commencing after 1975.	Lesser of gain on "farm recapture property" or balance in EDA. Only land deductions for 5 years on sale of land.
Recapture of Investment Credit. IRC Sec. 47 See § 518*	"Unearned" credit claimed on tangible personal property (includes livestock, other than horses) and other tangible property used in farming or for farm storage or research (except buildings) with useful life of 3 years or more.	Difference between credit claimed and credit based on period property held is added to tax.

*Rules applicable to investment credit in effect prior to April 18, 1969 are omitted. They do differ in some minor details.

Special Rules	Tax-Free Exchanges	Death, Gift, Other
Gain not recaptured is Sec. 1231 or capital gain. Subject to EDA recapture. See below.	Recapture limited to recognized gain plus value of nondepreciable property not treated as gain. Potential recapture carries over to new property.	No recapture on death; no pre-1977 carry-over. No income on gift but donee succeeds to recapture. Charitable gift reduced by recapturable depreciation.
"Applicable percentage" is 100% reduced by one percentage point for every month over 20 that property was held. No recapture if held over 10 years. Gain not recaptured is Sec. 1231 or capital gain. Post-1969 excess depreciation is recaptured first. Not subject to EDA recapture. See below.	Same as above.	Same as above. No recapture on residence except part used for business.
Post-1969 excess depreciation is recaptured before 1963-69 excess depreciation. See above. Residential rental units receive more liberal treatment. Gain not recaptured is Sec. 1231 or capital gain. Not subject to EDA recapture. See below.	Same as above.	Same as above.
Per cent of deductions recaptured if land held: 10 years none 9 years 20% 8 years 40% 7 years 60% 6 years 80% 5 years or less 100% EDA recapture applies first.	Same as above.	Same as above. Allocation on sale of part of land.
Farm recapture property is depreciable property, livestock, unharvested crops and land under Sec. 1231, but not buildings. EDA is increased by losses, reduced by farm income and amounts recaptured.	Same as above; except special rules for formation of corporation and partnership.	No recapture on death; no carry-over to estate. On gift, if potential gain on property given exceeds 25% of potential gain on property of donor, donee succeeds to proportionate share of donor's EDA.
Applies to casualties and thefts.	Recapture on exchange, involuntary conversion; not on mere change of business form.	Recapture on gift; not on death. Recapture if business use ceases.

chapter 6

TAX REPORTING AND RETURNS

Section

§ 6.1 CASH OR ACCRUAL BASIS

Farmers ordinarily use the cash method or the accrual method of reporting income, but a combination of methods may be proper if it clearly reflects income.

Cash-basis farmers report income when received, claim deductions when paid, and do not use inventories.

Accrual-basis farmers report income when it is earned, claim deductions when expenses are incurred, and adjust taxable income by the "paper" gain or loss in the inventory value of farm products at the end of the year.

Even though the accrual method is required for other taxpayers engaged in production and sale of goods, and even though the accrual method would more clearly reflect a farmer's income, most farmers may use the cash basis. A farmer who keeps no records or whose records are on the cash basis must report on the cash basis. However, after 1975 if a farm corporation ever has more than $1,000,000 in gross receipts in any year and if it is neither a family corporation nor one which has elected to have its income taxed to its shareholders, it must use the accrual basis. A partnership which has such a corporation as a partner must use the accrual method of accounting. They must also capitalize pre-productive expenses of property. A family corporation is one which is 50% or more owned by the taxpayer, ancestors, descendants, siblings, and spouses or estates of any of these.

Most farmers selected the cash or accrual method on the first return filed. Except where prohibited from using the cash method under these rules, a new election is made by a newly organized farm partnership or corporation and an election may be made for a newly purchased farm if it is a separate business. Once a method has been adopted, the farmer may not change it without the consent of the Commissioner.

The method of reporting for tax purposes must be the same as the method by which the farmer keeps his books. If he has no regular method or if his method does not clearly reflect income, computation of income will be made by a method that does clearly reflect income.

A farmer may report his farm business income on the cash basis and other business income on the accrual basis, or his farm business on the accrual basis and his personal affairs on the cash basis.

A corporation engaged in custom finishing livestock in feedlots is a "farmer" and can report on the cash basis even though the corporate books are on the accrual basis.

Cash Basis

Most farmers use the cash receipts and disbursements method of reporting income. This means that they can disregard inventories as a factor in computing income, account for income items when received, and claim deductions for expenses when paid.

Expense deductions for the cash-basis farmer are reported in the year when cash or its equivalent is paid out. An item is considered as paid in the year in which a check is mailed if the check is paid in due course.

Cash-basis reporting for the farmer (a) is simple; (b) reflects the farmer's actual cash position; and (c) lightens the tax load on sale of work, dairy, breeding or sporting animals raised because the entire sales price is capital gain (unless he has an "excess deductions account"). See § 5.13.

Before 1977, there was a tax advantage when a farmer on the cash basis died because his crops, livestock and other produce passed to his estate free of income tax, and the estate took the property with a tax basis equal to the fair market value at death. Neither the decedent nor his beneficiaries after his death paid the income tax which a farmer on the accrual basis would have paid upon the increase in value of the inventory.

Beginning in 1977, that advantage will disappear because these items will have the same basis to the estate or heirs as they had to the decedent, increased by any death taxes attributable to increases in value after 1976. This basis will also be increased for appreciation in value before 1977. See § 5.8.

Accrual Method

The accrual method of accounting reflects the income from ordinary farming operations more clearly than the cash method because the difference between the beginning and closing inventories of the farmer affects the reportable income or loss for the year. Costs which have gone into products not yet sold are, in effect, carried over to be offset against income when the sale is made.

The accrual method may not be used unless inventories are kept. An inventory is not adequate if made from memory or by estimate.

To determine gross income, the farmer on the accrual basis adds: (1) the sale price of all livestock and other products held for sale and sold during the year; (2) the inventory value of livestock and products on hand at the end of the year; (3) all miscellaneous items of income, such as breeding fees, rent of livestock, machinery or land, or other incidental farm income; (4) any subsidy or conservation payments that must be regarded as income; and (5) gross income from all other sources.

From this, he subtracts (1) the inventory value of the livestock and products on hand at the beginning of the year, and (2) the cost of livestock or products purchased during the year. The resulting figure is his gross income. Crop shares received as rents are not included in income until reduced to money or its equivalent or fed to livestock, nor are they inventoried.

The Service held in Revenue Ruling 76-242 that growing crops must be inventoried for taxable years beginning after June 27, 1976 (calendar year 1977 for most farmers) but it did not mention crop share rents. See § 3.4.

Expenses for which the farmer is obligated, whether paid or not, business losses and depreciation are deducted from gross income.

Livestock purchased for draft, breeding, dairy, or sporting purposes and not for sale may be included in the inventory, instead of being treated as capital assets subject to depreciation, if the practice is followed consistently from year to year.

Because the effect of accrual accounting is to reserve the costs of producing crops, livestock, or other products to set off against the price received when the items are sold, the result is a more even year-to-year taxable income.

The most serious tax disadvantage of the accrual method arises in connection with capital gains on the sale of draft, breeding, dairy, and sporting animals. If the animal is inventoried, the costs of raising are capitalized and the amount of the gain is only the difference between the last inventory value and the sales price.

Hybrid Method

A hybrid reporting system of any combination of methods is permitted if it clearly reflects income, for example, accrual of sales and purchases and cash reporting for other items. The method used for determining income must be used for determining deductions. If inventories are used, the accrual method must be used for purchases and sales unless special permission is obtained.

One should not confuse "hybrid" reporting with the right of the taxpayer with several businesses to use different accounting methods for separate business units. Separate books and records must be maintained for each business, and profits or losses may not be shifted by means of the different reporting systems. Also, a taxpayer may keep his personal records on a cash basis and his business records on an accrual basis.

Freedom from EDA Recapture

Before 1976, the decision to adopt or change to accrual accounting might have been influenced by the requirement that some cash-basis tax losses had to be kept in an "excess deductions account" which converts capital gain to ordinary income. However, after 1975, no additions to an EDA will be made, and it has ceased to be a factor in deciding the accounting method. See § 5.13.

Change of Reporting Method

Any change in the method of reporting requires the permission of the Commissioner of Internal Revenue. A change in reporting method means a change in the over-all method of accounting or a change in the treatment of any material item.

For practical purposes, a farmer may effectively change his method of reporting income by joining a partnership which elects a different method than his own. No consent of the Commissioner is required in making this "change."

Change from the accrual method to the cash method requires consent of the Commissioner which he has been reluctant to give because the usual reason is to obtain greater capital gains on breeding livestock.

Farm Income Tax Manual, § 102.

§ 6.2 CALENDAR OR FISCAL YEAR

A new taxpayer may elect a calendar or a fiscal year. Thereafter, the taxable year may not be changed without permission of the Commissioner.

A tax reporting year is selected by a taxpayer on the first tax return filed. The taxable year so chosen must be retained unless a change is authorized by the Commissioner. If a taxpayer does not elect a fiscal year, he must report on a calendar-year basis. Normally, a fiscal year must terminate on the last day of a month other than December. However, a fiscal year or a calendar year might end on any day of the month because of death of the farmer or the dissolution of a partnership, or the termination of a corporation.

A taxpayer may have only one reporting period. Income from all businesses and all other sources must be reported on the same taxable year. For example, while a farmer can report his farm business on the accrual basis and his other activities on the cash basis, he cannot report his farm on a fiscal year and his personal affairs or feed business on a calendar year. A new corporation, including a corporation taxed like a partnership, elects a fiscal year without reference to the year used by its stockholders, but a partnership must use the same year as its principal partners.

Farm Income Tax Manual, § 122.

§ 6.3 FARM INVENTORIES

Inventory valuation methods used by accrual-basis farmers are (1) cost; (2) cost or market, whichever is lower; (3) farm-price; and (4) unit-livestock-price.

All accrual-basis taxpayers may value inventories at cost, or at cost or market, whichever is lower, but farmers have options not available to other taxpayers to use the "farm-price" method, and a method applicable only to livestock called the "unit-livestock-price" method. An inventory valuation method once adopted must be followed consistently and may not be changed without the consent of the Commissioner of Internal Revenue, except that a farmer may change from cost, or cost or market to the unit-livestock-price method without consent. But the unit costs and animal classifications are subject to approval by the Internal Revenue Service.

All farm products held for sale must be inventoried. Supplies must be inventoried unless only small amounts are kept on hand. Crop shares received as rent are not included in inventory.

Growing crops must be included in inventory in taxable years beginning after June 27, 1976 (calendar year 1977 for most farmers). This is a change of a very old rule brought about by Revenue Ruling 76-242. See § 3.4.

"Cost" and "Cost or Market, Whichever Is Lower"

Although the cost method is a common way of valuing business inventories, this method is rarely used by small or medium size farmers. United States Treas. Regs., § 1.471-3 discusses the method in detail.

For large-scale farming operations, the difficulty in determining cost is probably not a sufficient objection to the use of the cost method of valuing inventories. It does accomplish the desirable objective of keeping closing inventory values low, so it is relatively better than the farm-price method in connection with sales of work, dairy, breeding, or sporting animals.

The "cost or market, whichever is lower" method of valuing inventories has advantages and disadvantages similar to those of the "cost" method, but is to be preferred in a period when market prices fall below cost. See United States Treas. Regs., § 1.471-4 for details.

Since, under the cost method, the closing inventory valuation (cost) can be higher than the market price and the taxpayer would have tax to pay on "book income" which might not materialize, there is obviously a tax advantage in being in a position to use a lower market value as a basis of valuation. If products are subject to wide price fluctuations, "cost or market" is generally preferable to the straight cost method.

Farm Price

Under the "farm-price" method, the inventory is valued at the market price less the direct cost of getting the product to market. The market price used is that of the nearest local market, not that at some distant point. No one but a farmer may use the farm-price method, but for the farmer its use is optional. He cannot be compelled to use it. A farmer may elect to inventory or capitalize draft, breeding, dairy, or sporting livestock under the farm-price method if he maintains a consistent practice.

If the farm-price method is adopted, it must be used for the entire inventory except for livestock which the taxpayer has elected to value under the unit-livestock-price method. A change from some other method to the "farm-price" method may not be made without the permission of the Commissioner of Internal Revenue.

Unit-Livestock-Price

The unit-livestock-price method is a substitute for a cost inventory. Different classes of animals in the inventory are valued at standard unit prices. This simplifies the problem of determining "costs" for inventory purposes. Since it is a cost inventory, the method does not react to market declines or increases.

Under the unit-livestock-price method, inventory items other than livestock, such as crops, must be valued by one or the other methods—(1) cost, (2) cost or market, whichever is lower, or (3) farm price.

The livestock raiser electing to use the unit-livestock-price method must adopt a reasonable classification of the animals in his inventory with respect to age and kind. The unit price assigned to each class should reasonably represent the average cost of producing an animal in that class.

All raised livestock held for draft, breeding, dairy, and sporting purposes must be included in the inventory if the farmer elects to use the unit-livestock-price method for his other animals. Purchased animals are included at cost. If immature animals are purchased, the cost is increased in accordance with the established unit price as the animal matures.

Farm Income Tax Manual, § 113.

§ 6.4 RETURNS WHICH MUST BE FILED

Single individuals under 65 with gross income of $2450 or more, and individuals 65 or over with gross income of $3200 or more must file income tax returns.

Married taxpayers' filing requirements are: $3600 if both are under 65, $4350 if one is over 65 and $5100 if both are 65 or over.

Married persons filing separate returns or living apart must file a return if gross income is over $750.

If self-employment income is $400 or over, a return must be filed even though gross income is less than the amounts above.

All partnerships and corporations must file returns.

Farmers may also be required to file declarations of estimated tax and information returns on amounts paid out, and to report

Social Security taxes paid on agricultural labor and income tax withheld.

Gas tax refund credits are claimed on the income tax return.

Form 1040 will be used by every individual farmer required to file a return and will be accompanied by Schedule F (the farm schedule), Schedule D, and Form 4797 to report gains and losses on sales of capital assets, assets used in the trade or business, draft, breeding, dairy, and sporting livestock, and other items, and Schedule SE to report self-employment income.

Every partnership must make a return on Form 1065 regardless of the amount of the partnership income or loss. A regular corporation files Form 1120, and a Subchapter S corporation files Form 1120S.

A farmer subject to Social Security tax must file an income tax return if his income from self-employment is $400 or more. This may include some farmers who would not otherwise have to file returns for income tax purposes, for example, a single farmer over 65 who has $3000 gross income but whose self-employment income is over $400. The receipt of Social Security benefits does not obviate this duty to report and pay tax on self-employment income. See § 7.3.

If two-thirds of his gross income comes from farming, a farmer must report and pay estimated tax by January 15 (calendar-year farmer) unless it is certain that the tax return will be filed by March 1. See § 6.5.

Other Reports

Farmers who pay $600 or more to individual workers, or who pay out that amount for rent, professional fees, interest or other payments that are income to the recipient, must file information returns on Forms 1099 and 1096. It is not necessary to report payments for merchandise, freight or similar charges or rent paid to real estate agents.

Farmers who employ agricultural labor during the year must report to the employee the amount of Social Security tax withheld. This may be done on Form W-2 or Form SS-14. A report must also be made to the District Director on Form 943. See § 7.2.

Voluntary Income Tax Withholding

An agricultural employer and employee may agree that the employer shall withhold federal income tax from the employee's wages. The withholding is reported on Form 943, Employer's Annual Tax Return for Agricultural Employees, at the same time Social Security taxes are reported. The employee must receive a Form W-2.

See Circular A, Agricultural, Employer's Tax Guide, for details.

Federal Gas Tax Credit

The federal gas tax refund is claimed as a credit on the farmer's federal income tax return. See § 6.6.

Extension of Time to File Returns

Farmers who need more time to file the Form 1040 may have an automatic extension of two months by filing Form 4868 before the due date of the return. A tentative estimate of the tax due and payment of the estimated balance due must be paid with the Form 4868. Interest at 7% is charged on the late payment.

Any further extension must be applied for in writing and be specifically granted.

This provision does not apply to extend the special March 1 filing date for farmers. To rely on the automatic extension, estimated tax must be paid on January 15 and the extension claimed from the April 15 due date of the return.

Farm Income Tax Manual, §§ 124, 126.

§ 6.5 DECLARATIONS OF ESTIMATED TAX

Farmers may file declarations of estimated tax annually (not quarterly) and may avoid declarations if the return is filed within two months after the close of the taxable year.

A farmer is an ordinary taxpayer in the sense that he can follow the regular declaration filing rules. He is a special taxpayer in that there are certain filing provisions for his benefit that are not available to other taxpayers.

A nonfarmer taxpayer files a declaration of estimated tax (Form 1040-ES) by April 15, and thereafter if there is any change in his estimate he files an amended declaration by June 15, September 15 and January 15. Payment is made in four or fewer installments. He can avoid the last January 15 declaration and payment if he files his return by January 31.

For farmers and other taxpayers, the self-employment tax is included in the estimate.

Special Rule for Farmers

A farmer is a person whose estimated gross income from farming is two-thirds or more of his gross income from all sources. Farmers are required to file declarations of estimated tax and make an advance payment but they have only one filing date. Their declarations must be filed by January 15 after the close of the calendar year. If a farmer files the declaration of estimated tax, his income tax return is due on April 15. However, a farmer has a special privilege of filing his income tax return by March 1 and avoiding the necessity of making a declaration.

For fiscal-year farmer taxpayers, the estimate is made by the fifteenth of the first month after the close of the taxable year unless the final return is filed by the end of the second month. Self-employed persons must include their estimated self-employment tax in their declaration of estimated tax.

What Is Farm Income

The term "estimated gross income from farming" includes all ordinary farm income plus the gains from the sale of livestock used in the trade or business of farming and held for draft, breeding, dairy, or sporting purposes. The entire amount of the gain from the sale of draft, breeding, dairy, and sporting livestock is treated as farm income; it is not reduced by the 50% deduction allowed with respect to such gains. "Gross income from farming" does not include gains from the sale of farm land or farm equipment, but these gains are included in gross income.

Landlords do not have "gross income from farming" for declaration purposes unless the rental income is based upon a percentage of farm production.

If a joint return is filed, two-thirds of the combined income must be "gross income from farming." In one case, the wife's income from teaching was 35% of the total gross income and all of the husband's income was farm income; they did not qualify and were subject to penalties.

Penalties

The payments of estimated tax must be 66 2/3% of the combined liability for income tax and self-employment tax to avoid penalty. The rule for nonfarmers is 80% of the tax due.

If there is an underpayment, the additional charge or penalty is figured at 7% per year on the underpayment from January 15 to the date of payment or to April 15, the due date of the final return, whichever is earlier. Form 2210F is to be used by farmers to determine if there was an underestimate and to compute the penalty.

If there is an underpayment, two common (and two other) exceptions may avoid the penalty. If the amount of tax actually paid, credited or withheld equals or exceeds the tax for the prior year, no penalty is due. Second, there is no penalty if the amount paid, credited or withheld equals or exceeds the tax computed on last year's income using the current year's exemptions and rates.

The easiest way to avoid the penalty for underpayment of estimated tax is to pay an amount equal to last year's tax. If there are four installments, pay one quarter of last year's tax in each installment. If a farmer makes a single January 15 declaration, he is free from penalty if he pays the amount of the prior year's tax.

There is no separate penalty for failure to file a declaration of estimated tax. The penalty for underpayment is based upon an actual payment of zero. However, one Revenue District, in effect, established a penalty for failure to file using an interesting argument. It said that the farmer's privilege to file a declaration on January 15 was an election. If the farmer did not file at all, he made no election, therefore he fell under the ordinary taxpayer rules requiring four declarations. It figured the penalty at 7% on the amounts that should have been declared on the April 15, June 15, September 15, and January 15 declaration dates. This would mean that a farmer who filed a declaration of $1 on January 15 would be liable for interest only from January 15 but if he did not file at all, the interest would run from the prior dates. The position seems wrong; but it points up the importance of filing some declaration on January 15. Undoubtedly other districts will make the argument.

Farm Income Tax Manual, § 128.

§ 6.6 FEDERAL GAS TAX CREDIT

The federal gas tax refund is claimed by a credit on the farmer's federal income tax return. A cash-basis farmer includes the refund in income in the year he files his return. An accrual-basis farmer treats the refund as income in the year he uses the gasoline.

An owner, operator, or tenant who buys gasoline for use on a farm for farming purposes may obtain a credit of four cents per gallon on his income tax return. A credit of six cents per gallon is allowed on lubricating oil used otherwise than in a highway motor vehicle.

The person who actually bears the cost is entitled to the refund. Thus it could be an owner, operator, partnership or corporation which makes the claim. A landlord or tenant might have borne all or part of such cost.

The refund credit is claimed on his income tax return by the farmer for gasoline used by him or by someone else on his farm for farming purposes. Where custom work is done, the farm owner, tenant or operator claims the credit; the custom operator does not. If a landlord and tenant share gasoline cost, each claims his own credit on the gasoline he paid for. If a farmer sells gasoline to a neighbor, the neighbor claims the credit.

Generally, the amount of the refund must be claimed as a credit on the income tax return. However, if $1000 or more is payable with respect to gasoline used during any of the first three quarters of the taxable year a claim for refund may be filed with respect to the gasoline used during such quarter. The refund for the last quarter may be claimed only as a credit on the income tax return.

Form 4136, Computation of Credit for Federal Tax on Gasoline and Lubricating Oil, is to be used to claim the credit on the federal income tax return for the excise tax (1) on gasoline used on a farm for farming purposes, and (2) on lubricating oil used otherwise than in a highway motor vehicle.

A cash-basis farmer deducts the full cost of the gasoline used on the farm and reports the credit as income in the year in which he files the income tax return on which he claims the credit. An accrual-basis farmer technically accrues his right to the refund as he uses the gasoline, therefore he reports the amount of credit as income (or a reduction of his gasoline cost) in the year in which the gasoline is actually used.

Farm Income Tax Manual, § 310.

§ 6.7 ACCOUNT NUMBERS

Every taxpayer, individual or other, must have an account number which will appear on tax returns, statements or documents. Generally, an individual uses his Social Security Number and other taxpayers use an Employer Identification Number.

An employer must have an Employer Identification Number.

Account numbers must be shown on every income tax return, statement, or document. An individual taxpayer's account number is his Social Security Number. If a husband and wife file a joint return or declaration, the account number of the husband only is shown unless the wife has gross income of her own, or self-employment income, or if she receives payments such as dividends and interest which must be reported on an information return. On separate returns the account numbers of both husband and wife should be shown on each return.

On all business returns, the account number is the Employer Identification Number. Every employer (except a household employer) who has one or more employees subject to Social Security or income tax withholding must have an Employer Identification Number.

A sole proprietor with no employees uses his Social Security Number but a corporation, partnership, trust or estate which has no employees must have an Employer Identification Number. An individual proprietorship which is incorporated is a new legal entity and it must obtain a new Employer Identification Number.

An individual engaged in a trade or business will show his Social Security Number on his return, and both his Social Security Number and his Employer Identification Number, if any, on the Business or Farm Schedule.

Payments of Dividends, Interest, etc.

Payors of dividends and interest or others who file information returns with the Internal Revenue Service must include the payee account numbers on these returns. Failure to include this number can bring a $5 penalty for each default. This penalty has not been enforced in the past, but the Service has announced that the penalty will be enforced after January 1, 1977.

Even though a person is not required to file an income tax return, if he is a payee for whom an information return must be made, he must have a number. For example, a child who receives dividends must give a number to the payor even though the child does not have to file an income tax return.

Use the 1099 Figure

Taxpayers should ordinarily report on their tax returns the same amounts of interest and dividends shown on the statements they receive from the paying organizations. A taxpayer who thinks the statement is incorrect should contact the paying organization. The IRS finds that the discrepancies seriously upset its computers.

Farm Income Tax Manual, § 129.

§ 6.8 AVERAGING OF INCOME OVER FIVE YEARS

A farmer whose income in any year is one-fifth greater than his average income for the prior four years may elect to have the excess income taxed at a lower than normal rate.

If a farmer's income in any year is in excess of 120% of his average income for the prior four years (and the excess is also over $3000), all the excess income will be taxed at the rate that applies to the first one-fifth of the excess income.

There are two basic ideas in this "averaging" scheme. First, the "averageable income" of the current year (called the "computation year") is the amount of taxable income over 120% of the average taxable income of the past four years (called the "average base period income"). For example, if a farmer had taxable income of $14,000 in 1976, $9000 in 1975, $10,000 in 1974 and $7000 in 1973, the average would be $10,000. The amount of "averageable income" in 1977 is the excess over $12,000, that is 120% of the $10,000 four-year average. But the averageable income must also be over $3000. Thus, 1977 income must be in excess of $15,000 for the averaging provisions to apply.

The second basic idea is this: the tax on all of the excess or "averageable income" is five times the tax on the first one-fifth of the excess income. In effect, the progressive rate schedule is eliminated on the top four-fifths of the excess income. For example, if the farmer in the above example had taxable income of $22,000 in 1977, his averageable income would be $10,000, the excess over $12,000. The tax is computed in the regular manner on the base income of $12,000 plus one-fifth of the excess income ($2000), or $14,000. The amount of tax on the $2000 at the $12,000 to $14,000 bracket is multiplied by five. That is the tax on the $10,000 of averageable income which is added to the tax on the base income of $12,000.

Adjustments to Taxable Income

There is no adjustment of the current year's income or the income in the base period years for long-term capital gains. If capital gains caused the current increase in income, the taxpayer obtains the usual 50% deduction and the averaging provisions will lower the rate on the taxable half of the gains.

Items not eligible for averaging are certain throwback distributions from trusts and penalty distributions from a self-employed retirement plan. Excludable income from sources outside the United States must be added back. Adjustments for these items are also made in determining income in the base years.

Special Rules

Averaging is an election; it is not mandatory. If an election to average is made, the taxpayer may not use the optional tax table nor the 50% maximum rate on earned income nor the alternative capital gains tax computation.

Averaging may be used only by a citizen or resident subject to the tax during the full five-year period. A provision that a taxpayer must have furnished more than half of his own support prevents dependent students from averaging the incomes from their first full-time jobs with the prior low-income school years, but it does not deny averaging to a person who was unemployed in prior years, or to a person whose current income is the result of work performed in prior years, or to a self-supporting person because he or she marries a person who was not self-supporting.

If a husband and wife filed joint returns in some years and separate returns in others or were not married during the five-year period, the income is reconstructed to adjust for these factors.

Farm Income Tax Manual, § 123.

chapter 7

SOCIAL SECURITY TAXES AND BENEFITS

Section

§ 7.1 AGRICULTURAL COVERAGE

People in farm activities are generally covered by Social Security if their earnings are from "employment" or "self-employment." An employed person works for someone else. A self-employed person works for himself. While the benefits are similar, the methods of taxing employed and self-employed people are different.

A farm employee subject to Social Security pays a tax on his wages. His employer pays a like tax on the employee's wages. A self-employed farmer pays his tax on his "net earnings from self-employment."

(a) Agricultural labor

A hired farm worker is an employee. He is generally paid a fixed amount for his labor and is under the supervision of his employer.

Board and lodging, share of crops and payment in grain or livestock are not treated as cash wages. A cash wage test is applied to wages paid by each employer to each employee.

A farm worker is covered by Social Security and he and his employer are subject to tax if either one of two tests is met:

(1) He received cash wages of at least $150 from any one farm employer during the calendar year, or

(2) He works for any one farm employer on twenty or more days for cash remuneration computed on a time basis (hourly, daily, weekly).

Thus it is possible that a farm employer will have to withhold tax and pay tax on wages paid to a worker even though total payments are less than $150. It is important to keep track of days and parts of days worked by each employee.

Payment of the employee's share of the tax by an employer who does not deduct it from the employee's wages is not treated as additional wages for additional Social Security tax.

Practically all farm workers are covered by the law. However, there are certain family situations in which no tax is due and no benefits result. These are not covered:

(1) A spouse who works for the other spouse.

(2) A parent who works for his or her child unless the parent works in the child's trade or business.

(3) A child under 21 who works for his or her parent.

"Agricultural labor" generally includes all services performed on a farm in the employ of any person in connection with cultivation of the soil or raising commodities or livestock, and services in the employ of the owner, tenant or operator in connection with farm operation and maintenance if the major portion of the service is performed on the farm.

Certain other services included in the term involve cotton ginning, maintenance of waterways, and processing activities where the processor produces half of the commodity. Domestic workers on farms operated for profit are treated as farm employees.

What Is Employment?

For purposes of determining whether a person is an employee, the Social Security law turns to legal definitions developed over many years. These rules are not always sharp and clear, but certain basic ideas can be found. The distinction is normally made between an employee (originally thought of as a servant in a master-servant relationship) and an independent contractor. An employee performs services for another, subject to the control of the other, his employer. This does not mean that the employer must stand by, constantly observing and supervising. It does mean that an employer has the right, whether he actually exercises it or not, to have the work performed in such manner and detail as he directs, and the employee has a duty to perform the work as directed. A hired hand, a sales clerk, a worker in a feed mill, and a bus driver are employees.

An independent contractor is one who agrees to perform certain services for a client, customer or constituent. His function is to do a job for a price, the finished job to meet certain specifications. The manner of doing the job and the control over methods and details is retained by the contractor. The television or car repairer, the lawyer, and the person who does custom work are independent contractors.

Lines are often difficult to draw. Employees may go to great lengths without actual supervision, while the repairman or the custom baler may have to put up with considerable supervision. The determination depends upon the particular facts in each case.

Crew Leaders

In some parts of the country, workers are made available to the farmer by a "labor contractor" or "crew leader." If the crew leader is the employee of the farmer, the contractor and all of the workers are agricultural laborers and the tax is levied upon each employee and the farmer employer.

The crew leader will be the employer of the crew in most instances. If the crew leader furnishes individuals to do farm labor for a farmer and pays them for their work either on his own behalf or on behalf of the farmer, the crew leader, not the farmer, is the employer of the crew members unless the crew leader has entered into a written agreement with the farmer in which the crew leader is designated an employee of the farmer.

(b) Self-employed farmers

Self-employed farmers are subject to the "self-employment" tax, which is based upon net earnings from self-employment in the farm business. The tax is computed on Schedule SE (Form 1040) and paid when the farmer files his income tax return.

A child is self-employed if he carries on an activity of his own, such as a 4-H project, and has net earnings of $400 or more.

Farm Income Tax Manual, §§ 1000, 1002.

§ 7.2 EMPLOYERS OF AGRICULTURAL LABOR

The employer must have an identification number, keep records, withhold tax from the employee's wages, make reports, and pay the tax levied upon employer and employee.

Each farm employer must have an Employer's Identification Number which is used on all forms reporting the employer's Social Security tax payments.

An employer must keep the following records for each employee to whom he pays or expects to pay (1) cash wages of $150 or more in the year, or (2) cash wages for work on a time basis on twenty or more days in the year:

1. Name and Social Security account number.
2. Cash wage payments to the employee.
3. Daily records of time worked and basis of payment.
4. Amount deducted as employee tax.

The employer must make reports and pay taxes on each employee who met either of the two wage tests in the year. The tax for each year is shown in the table below. The tax applies to the first $15,300 in 1976 and $16,500 in 1977. Thereafter, the wage base will be fixed by a formula that follows the growth in nationwide wages.

Tax Rates for Employer and Employee

Year	Social Security Tax Rate	Hospital Insurance Rate	Total
1976-77	4.95%	0.90%	5.85%
1978-80	4.95%	1.10%	6.05%
1981-85	4.95%	1.35%	6.30%
1986-2010	4.95%	1.50%	6.45%
After 2010	5.95%	1.50%	7.45%

The employer deducts the employee's part of the tax when he pays the wages to the employee. The employer has the right to deduct the employee tax even before the employee has received $150 or worked on twenty separate days in the year. If he prefers, he can wait and deduct the tax when the qualifying limit is reached.

If the employer deducts from the start and it turns out that the employee does not earn as much as $150 or does not work on twenty days, the employer should repay the employee if he can. If he cannot repay the employee, he should send the deducted amount with the name of the employee and an explanation to the District Director of Internal Revenue in January after the year in which the amount was deducted.

In order to get his refund the employee must apply to the Internal Revenue Service or claim the excess Social Security tax as a payment on his income tax.

Every employer who pays taxable wages for agricultural labor must make a return on Form 943 (Employer's Tax and Information Return for Agricultural Employees) before January 31 after the year in which the wages are paid.

Government Depositaries

The cash amount of taxes (employee and employer share) for agricultural labor must be deposited in a government depositary. The rules, forms and instructions may be obtained from the District Director of Internal Revenue. The local bank may be an authorized depositary and have the information.

Statement to Employee

An employer must provide each employee with a written statement of his taxable wages and employee tax for each year on or before January 31 of the next year. If the employment ends before the end of the year, the statement should be provided within thirty days of the last wage payment.

Form W-2 or Form SS-14 should be used. Include on the statement the name and address of the employer, the name, address and Social Security number of the employee, the total amount of wages subject to Social Security tax paid during the year, the amount of Social Security tax deducted from the employee's wages, and the amount of tax attributable to hospital benefits.

Farm Income Tax Manual, § 1003.

§ 7.3 SELF-EMPLOYED FARMERS

Every self-employed farmer must have a Social Security card and Social Security account number. He computes his self-employment tax on Schedule SE (Form 1040) and pays it with his income tax.

The self-employed farmer pays his self-employment tax at the time that he files his income tax return. Schedule SE (Form 1040) provides space for computing self-employment income and Social Security tax. The amount of self-employment income subject to tax is reduced by any wages received which were subject to Social Security taxes. The farmer may owe a self-employment tax even though he owes no income tax.

No tax is due if self-employment income is less than $400. If self-employment income is $400 or over, the tax is based on all such income up to $15,300 in 1976 and $16,500 in 1977 (less the amount of any covered wages). Thereafter, the self-employed base will be determined by a formula that follows the growth in nationwide wages.

See the special option for farmers to compute self-employment income on gross income in § 7.8.

Self-Employment Tax Rates

Year	Social Security Tax Rate	Hospital Tax Rate	Total
1976-77	7%	0.90%	7.90%
1978-80	7%	1.10%	8.10%
1981-85	7%	1.35%	8.35%
After 1985	7%	1.50%	8.50%

If a self-employed person reports his earnings on a fiscal year basis, he uses the rates that apply to the calendar year in which his fiscal year began.

Farm Income Tax Manual, § 1004.

§ 7.4 FARM OPERATORS AND PARTNERS

Farm operators, tenants and members of farm partnerships are self-employed. Farm landlords are not self-employed unless they have leases which provide for material participation in the farm operation and they do, in fact, materially participate.

Usually it is easy for a farmer to tell if he is a self-employed person subject to Social Security tax and eligible for Social Security benefits. Here are the common types of farm activities and the status of each as self-employment.

(a) Farm owner-operators

Farm owner-operators are covered by the Social Security law as self-employed farmers. To be self-employed, a farmer must operate the farm himself, or use hired labor or be a member of a partnership that operates the farm.

Farmers operating under various forms of contractual arrangements for production of farm products are usually treated as self-employed, not as employees of the contractors.

(b) Tenants and share farmers

A farmer who rents land from another for cash or on a sharing arrangement is self-employed and is covered by the law. There are many variations of tenancy or sharing. They range all the way from a simple crop-sharing with little or no contribution to an arrangement in which the tenant is practically a partner with the landlord. Since share-farmer, tenant and partner are generally treated as self-employed, the person who farms someone else's land is almost always covered by the Social Security law.

(c) Landlords

The fundamental rule is that a landowner who has leased his land to a tenant or who has his land in some form of crop-share arrangement is not self-employed for purposes of liability for self-employment taxes or qualification for Social Security benefits. It does not matter whether he has a cash lease, a crop-share lease, or a livestock-share lease. Rentals, including rents received in crop shares, are not self-employment income.

However, in recognition of the fact that many share leases are virtually partnerships, the law provides that a landlord's income is self-employment income if certain tests of material participation are met. Because of the importance of the material participation lease and the detail of the rulings and interpretations, problems relating to these matters are discussed in a separate section. See § 7.5.

(d) Farm partners

If a partnership operates a farm, each of the partners is credited with a share of the self-employment income of the partnership. This means that it is the type of income of the partnership that is the determining factor. It does not matter whether a partner is actively engaged in the farm operation or not.

If a farm partnership held land which it rented to a tenant or turned over to a share-farmer for a share of the crops, the income would be treated as rent, not as self-employment income. If one or more partners "materially participate" in the farm operation, the partnership income will be self-employment income to all partners. If the partnership operates the farm business, raising and selling the crops, partnership income will be self-employment income and each partner will report his share of the income.

The activity or lack of activity of the individual partner has nothing to do with the determination of whether the partnership produces self-employment income. The partnership is the business unit and the partners share whatever income the partnership produces.

A limited partner in a limited partnership under the Uniform Limited Partnership Act reports his share of partnership self-employment income even though by law he cannot render services or engage in management without becoming a general partner.

Generally, if both husband and wife work on the farm, only the one who has the management and control will be regarded as receiving self-employment income. The fact that the land is held in co-ownership with the wife and the fact that she carries the usual load of an active farm wife will not make her a partner. A wife can be a partner if a valid partnership is established between husband and wife or among them and others.

Father-son arrangements have endless variations. Father and son could be true partners, both self-employed. Generally a sharing arrangement, labor share or tenant share, will produce self-employment income only to the laborer or tenant, not the landowner. But many father and son sharing arrangements fall within the terms of the "material participation" provisions making the father self-employed, too. See § 7.5.

(e) Combinations of farm operations

It is possible that a single individual may be engaged in several farm activities some of which are covered by Social Security and some of which are not. Income from the covered activities is subject to tax and produces benefits; other income does not. For example, a farm owner may rent part of his farm and operate part or he may have two farms, one rented and one personally operated; also he might be a self-employed tenant on another farm, and rent out his owned farm. Even though his whole farm is rented to a tenant, a farm landlord may have a separate livestock operation or a partnership livestock operation with the person who rents his land.

Farm Income Tax Manual, §§ 1005 to 1007, 1009, 1010.

§ 7.5 FARM LANDLORD COVERAGE

The most difficult problem in farmers' Social Security has been the status of the farm landlord. Under the "lease" umbrella are arrangements that range from mere share-cropping of land to virtual partnerships. The farm landlord's activities determine whether he has self-employment income and whether he has income that will reduce his benefits.

The landowner or landlord has no option to be covered or not covered by Social Security. The law is compulsory; self-employment income is subject to tax and produces benefits. Because certain types of farming operations produce self-employment income and others do not, a form of choice arises. The choice lies in the method of doing business.

Landowner's Choice

A landowner may be self-employed in four ways: (1) he may personally cultivate the farm using such hired labor and custom work as he needs; (2) he may personally operate the farm business using hired labor, which can include a hired manager; (3) he may set up a partnership to operate the farm, using personal labor of some or all of the partners or using hired labor; or (4) he may "materially participate" in farm operations under a lease or other share arrangement.

The self-employed status of farm operators and partners is discussed in § 7.4.

Material Participation

Payments in cash or kind received by a landowner under a lease or other arrangement with a tenant or share farmer will be self-employment income:

(1) If the arrangement provides for material participation by the landowner in the production or management of the production of the agricultural products, and

(2) If there is in fact material participation by the landowner in the production of the agricultural products.

The arrangement, that is, the written or oral agreement, must specifically provide that the landowner will materially participate in production or management and that the tenant shall produce agricultural commodities on the landlord's farm.

The arrangement need not be in writing, but existence or nonexistence of material participation is easier to prove from a written agreement. An oral understanding is shown by the landlord's acts, the tenant's acts, how both acted in previous years on similar matters, common practices in the area and statements of the landlord, tenant and third parties who may know the facts.

In addition, there must be actual participation by the landlord in production or management to a degree which is material. "Material participation" may be established by taking part in the work of producing the commodity or by taking part in management of production. It does not require both. A combination of physical work and management decisions can be a material contribution. Although "no hard and fast rules can be laid down" to determine the degree of participation, the factors involved in material participation are pretty well agreed upon.

Social Security people look to the amount of actual work in producing commodities, the extent of participation in management decisions and the importance of these decisions to the success of the enterprise, the frequency and regularity of consultation with the tenant about production, the frequency and regularity of inspections of production activities, the amount of tools, equipment and livestock furnished, and the amount of production expenses advanced, paid or otherwise obligated for. These must be required by the arrangement. Voluntary activities do not count.

Physical work and management decisions are the principal factors. Furnishing the machinery, equipment and livestock, or furnishing or advancing funds, or assuming responsibility for expenses are additional factors which may be considered in arriving at a decision in a borderline case.

Participation by an Agent

Before 1974, the activities of an agent, including a professional farm management company, were attributable to the landlord in determining whether the landlord materially participated. The law was changed, effective in 1974, so that it is no longer possible to materially participate through an agent.

Presently, material participation can arise only through the activities of the landowner or his employees.

The Owner Drawing Benefits

This section has emphasized the creation of self-employment income. Some forms of material participation may lose benefits after qualification. The farmer aged 62 to 72 years may desire to avoid participation because it will reduce his Social Security benefits. See § 7.7.

Although the discussion in this section has been phrased in terms of landlord or landowner for convenience, it applies with equal force to a tenant, partner or other person entitled to possession who subleases or has a similar share-farming arrangement.

Farm Income Tax Manual, § 1008.

§ 7.6 SELF-EMPLOYMENT INCOME

For most farmers, self-employment income will be the net farm profit shown on Schedule F (Form 1040). The tax in theory falls upon income from personal services, so there is eliminated from self-employment earnings investment type income and gains and losses from sales of assets used in the business.

Specifically excluded from net farm profit to obtain self-employment income are:

(a) All rentals from real estate and from personal property leased with the real estate (including crop-share rents) and all deductions related to such rentals.

(b) Income from dividends, from interest on corporate and government bonds in registered form or with coupons attached and from nonbusiness interest.

(c) Gain or loss from cutting of timber or disposal of timber or coal or iron ore entitled to capital gain treatment.

(d) Deduction for a net operating loss.

(e) Gains and losses from the sale, exchange or involuntary conversion of any asset that is not:

 (i) stock in trade or inventory property, or

 (ii) property held primarily for sale to customers in the ordinary course of the farm business.

This requires subtracting gains and adding back any losses that result from capital assets, draft, breeding, dairy, and sporting animals, farm equipment, and any other asset used in the business, regardless of the period held, except goods held for sale. Personal deductions, low income allowances, standard deductions and deductions for personal exemptions do not affect the computation of self-employment income.

Income from Operations

Thus, self-employment income works out to be income from regular farm operations. It is based upon the computation of farm income for income tax purposes with the above modifications. It includes patronage dividends, prizes, refunds such as gasoline tax, crop-damage payments, merchandise received for produce, and all other ordinary farm income items. It excludes Section 1231 items such as timber royalties, gain on cutting timber, crops sold with the land, sales of breeding animals, etc. Other factors in determining taxable income, such as the use of the cash or accrual basis, election to expense soil and water conservation expenditures, election to treat CCC loans as income, etc., are controlling for determining self-employment income.

Conservation reserve program payments are self-employment income even though the whole farm is in the conservation reserve. However, if the farm is rented, then the Government payments follow the classification of the rent. If the landlord materially participates (or did, if the whole farm is in), the payments to him are self-employment income. If he does not (or did not) the payments are rent.

A lump-sum payment (in lieu of annual payments) received under the Cropland Adjustment Program is treated as self-employment income in the year received by a cash-basis farmer and is not prorated over the period of the contract. The same result should follow if the farmer is on the accrual basis.

Crop shares are rents and are not self-employment income unless the landlord is materially participating in the farm operation. The crop shares are treated as income when reduced to money, fed to livestock or donated to charity. See § 3.4.

Grain storage fees paid to a farm operator under the price support loan program of the Commodity Credit Corporation are self-employment income but not to a landlord unless he materially participates.

The wheat, cotton and feed grain program payments are net earnings from self-employment. In the case of a landlord and tenant diverting land from production of wheat or feed grains, the landlord's share of payments is includable in computing net earnings from self-employment only if he materially participates in the production or management of the production of the farm commodities produced on the part of the land that is not diverted. If the landlord does not materially participate, then his share of the income under the programs is rent and not net earnings from self-employment.

Income received from the sale of gravel which is treated as a sale of a capital asset is not received in the ordinary course of a trade or business and is not includable in net earnings from self-employment.

Capital gain timber sales are not self-employment income. Sales of timber from farm wood lots reported as ordinary income are self-employment income earned from the business of farming.

Trade or Business

Determination of self-employment income turns not only on the nature of the income itself—it must also be produced by a trade or business. Any farm operator, tenant, share-farmer or any material participator is engaged in the trade or business of farming. The existence of a trade or business depends upon facts showing the regularity and the extent of the operation. Generally the operation is a trade or business if the person engages in it regularly for profit and it constitutes, in whole or in part, his livelihood.

A trade or business does not have to be a full-time activity or even have to do very well. For example, a rancher who developed his ranch into a hunting preserve on which he sold hunting rights on a per day, per week, or per season basis had self-employment income from a trade or business.

Farm Partners

A partnership computes self-employment income as an entity. Self-employment income is allocated to each partner on the partnership return.

Distributive share and guaranteed payments received from a farm partnership are reported by the partner for self-employment purposes (or added to other self-employment income) on Schedule SE. If the gross income option is used, the share of partnership gross income is aggregated with gross income of other farm activities, if any.

Even an inactive, silent or limited partner reports his share of self-employment income from the trade or business of the partnership. See § 7.4.

The income share of a deceased partner up to the date of death is treated as self-employment income. But, this has no effect on whether the deceased partner's distributive share of partnership income is reported in his final return for income tax purposes. This means that the self-employment income to the date of death of the partner-decedent must be separately computed unless the partnership taxable year did in fact end on the date of the decedent's death. See § 8.8.

Farm Income Tax Manual, §§ 1011 to 1013, 1015, 1016.

§ 7.7 TAXES VS. BENEFITS UPON RETIREMENT

After retirement, self-employment income continues to create a Social Security tax liability. The self-employment income can be derived only through personal participation in the farm activity. If a farmer performs substantial services in self-employment after retirement, the income will reduce his Social Security benefits.

If a farmer wants to retire from the active operation of the farm and turn the operation over to a family member or a farm tenant, he will be concerned: (1) whether the income derived from the rental arrangement continues to be self-employment income subject to tax, and (2) whether he is performing substantial services in self-employment which will reduce benefits. The self-employment income can be created only by the landowner personally participating in the operation of the farm. Activities of an agent, such as a professional farm-management company, are not attributable to the landowner so as to make him self-employed.

Social Security taxes have increased and are scheduled to go even higher; the self-employment tax payment on $16,500 in 1977 will be $1303.50. To avoid Social Security taxes the farm income must be rental income.

Avoiding Landowner Self-Employment Income

For the landowner to avoid self-employment income, the rental arrangement and his acts must not fall within the material participation tests. See § 7.5.

A direct lease arrangement with a tenant should provide that the operation of the farm and the principal management decisions are left to the tenant. In such cases the landlord would not be considered as materially participating in the operation of the farm and the rental income would not be considered self-employment income.

Since 1974, no problem arises where the landowner has turned over the operation of the farm to an agent farm manager. The agent may have control over planning the farm operation and participate in the management decisions. Before the change of law, these factors could result in material participation by the landowner through his agent.

The landowner may have the benefit of a farm management service without the cost of the self-employment tax.

Avoiding Reduction of Benefits

In 1976, a retiree may lose Social Security benefits if he or she earns in excess of $2760 a year from a job or from substantial services rendered in self-employment. If the farm rental is considered as income from substantial services rendered in self-employment or partnership business, amounts over $2760 a year will reduce Social Security benefits. After age 72, any amount may be earned without reduction of benefits.

For 1977, there is no reduction of benefits if the beneficiary earns no more than $3000 for the year or $250 per month. Amounts received in excess thereof reduce benefits. Each dollar over the limit reduces benefits by fifty cents.

In future years, the amount of allowable earnings will rise according to a formula tied to the increases in the wage base.

Substantial services rendered in self-employment are measured by a different test than material participation for self-employment income. In essence, the test for substantial services is whether or not the retiree is, in fact, retired. Substantial services arise only from personal participation in the farm activities. The Social Security regulations spell out various factors which are used to determine whether or not a landowner has rendered substantial services. Some of the factors to be considered are:

(1) The amount of time devoted to the operation.

(2) The nature of the services.

(3) The relationship of the work performed before retirement and after retirement.

(4) Whether there is a paid manager, a partner, or a family member who manages the business.

(5) The type of business.

(6) The amount of capital invested.

(7) The seasonal nature of the business.

The regulations say that 45 hours of service in a month is usually regarded as being substantial, though this is not a hard-and-fast rule. If the operation of the farm or ranch is turned over to a tenant and/or to a professional farm or ranch manager, the income does not result from substantial services rendered in self-employment, and thus does not reduce Social Security benefits.

Maintaining Benefits

There are many court decisions relating to attempted arrangements to pay a "retired" person a monthly income within the earnings limitations so that he can continue to draw full benefits and still run the business. Most fancy arrangements have failed. However, there are bona fide arrangements that work and are fully within the law. For example, a farmer could retire from the active operation of the farm and work for his son, relative, or even tenant on a limited basis and receive a salary commensurate with the services rendered. However, to stay within the earnings requirement, the work may not be on a full-time basis; he might work a limited number of days a week or on a seasonal basis for a salary under the yearly earnings ceiling or monthly exclusion.

However, if the retiree purports to rent the farm to a tenant or relative and spends as much time in the operation and management of the farm as he did before his supposed retirement, the arrangement will probably be considered a sham and the owner will be treated as performing substantial services in self-employment. All of the rental income would then be considered income which would reduce Social Security benefits.

Farm Income Tax Manual, § 1023.

§ 7.8 FARMERS' GROSS INCOME OPTION

Farmers have a special option to figure self-employment income on gross income if "net earnings" from self-employment are low. Farmers can get credit for $400 to $1600 of self-employment income without any farm profit.

A farmer may compute his net earnings from self-employment in the regular manner or he may use an optional method based upon gross income to compute earnings from self-employment. In effect, this guarantees some self-employment income in cases where the farming operation results in a loss or an extremely low net earnings figure.

Gross income means the gross receipts of the farm business, reduced by the cost of items purchased and sold in the business and adjusted for the income, deductions and gains and losses that are not treated as affecting self-employment income. See § 7.6.

These are the option rules:

(1) If gross income is $2400 or less, a farmer may report two-thirds of gross income as self-employment income.

(2) If gross income is more than $2400, and net earnings from self-employment are less than $1600, a farmer may report $1600 as self-employment income.

The choice of using actual self-employment income or the optional method may be made each year by the farmer. Even though he has other earnings or self-employment income, he may still use either the regular or optional method in figuring farm self-employment income to be added to other covered earnings or income.

Accrual-basis taxpayers, as well as cash-basis taxpayers, may use the option. Members of farm partnerships may use the option on their prorata share of partnership gross income. If the farmer has gross income from his own operation and from a partnership, he aggregates the amounts.

Farm Income Tax Manual, §§ 1014, 1015.

§ 7.9 SOCIAL SECURITY BENEFITS AND MEDICARE

Insurance benefits provide: (1) Retirement payments (a) to a retired farmer at age 62, (b) to his wife at age 62 or if she is caring for a child entitled to benefits, and (c) to certain children; (2) payments because of disability to disabled workers and their dependents and to disabled children over eighteen; (3) survivor payments which include a death benefit and monthly payments (a) to a widow at age 60 or caring for a child, (b) to children under eighteen, and (c) to certain widowers and dependent parents.

Most persons 65 and over are eligible for hospital insurance and voluntary medical insurance which pays hospital expenses and doctors' bills.

(a) Generally

There are three kinds of Social Security benefits: (1) old age and survivors' benefits, (2) disability benefits, and (3) hospital and medical benefits.

Retirement and survivor benefits stem from a fully insured status. Some survivor benefits arise from a currently insured status. Disability benefits turn on a fully insured status plus other requirements. Persons over 65 who are qualified for retirement benefits are covered by the hospital insurance provisions. Insurance which pays doctors' bills is voluntary and is obtained by the payment of monthly premiums.

Currently Insured

Currently insured status provides death and some survivor benefits. A person is currently insured if he had at least six quarters of coverage in the last thirteen quarters of his life. The quarter in which he died is the thirteenth. Any six of the previous twelve quarters must be covered. Currently insured status is also obtained if there are six quarters of coverage in the thirteen quarters, ending with the quarter in which a person became entitled to disability benefits or retirement benefits.

Fully Insured

Fully insured status provides retirement benefits. To be fully insured a person needs one quarter of coverage for **each year** which has elapsed since 1950 (or after the year in which he became 21 if that was after 1950) and before age 62.

"Quarters of coverage" is the measuring factor for eligibility for Social Security benefits; it has no effect upon the amount of benefits. A covered quarter is one in which a person has the minimum earnings required to obtain credit toward insured status.

Quarters needed to qualify to become fully insured are shown in the table below. The year is the year of death or the year of attainment of age 62.

Age 62 in:	Quarters needed:
1975	24
1976	25
1977	26
1978	27
1979	28
1980	29
1981	30
1986	35
1991 or later	40

Prior to 1975, the qualifying rules were based on the year a male reached 65 and a female reached 62. As of 1975, the qualifying rules for men are based on the year the claimant reaches 62, as is the case with women retirees.

Insured for Disability

A special test of coverage applies for disability benefits. In general, to be eligible for disability insurance benefits, a person must be "fully insured" (in the sense that he would have qualified for retirement benefits if he had attained retirement age and filed application) and must have at least twenty quarters of coverage in the previous forty quarters. If disabled before age 31, there is a special rule—half of the quarters since age 21 must be quarters of coverage.

Types of Benefits

Benefits of different types accrue to persons who have different insured classifications. "Currently insured" status provides a burial benefit and limited survivor benefits. Fully insured status produces virtually all of the retirement and death benefits. Benefits for a dependent husband or widower require both fully and currently insured status. A disabled person who qualifies is entitled to the same benefits for himself, his wife and his dependents as a fully insured person.

For quick comparison, here in chart form are the benefits that flow from currently insured, fully insured status, and both.

BENEFITS	Fully Insured	Currently Insured	Both
Retirement benefits			
Retired worker at 62	Yes	No	Yes
Disabled worker under 65	No	No	Yes
Wife of retiree at 62	Yes	No	Yes
Wife caring for child	Yes	No	Yes
Divorced wife at 62	Yes	No	Yes
Unmarried child or grandchild under 18 or student or disabled	Yes	No	Yes
Dependent husband at 62	Yes	No	Yes
Survivor benefits			
Widow at 60	Yes	No	Yes
Disabled widow at 50	Yes	No	Yes
Surviving divorced wife at 60	Yes	No	Yes
Disabled divorced wife at 50	Yes	No	Yes
Widow or widower caring for child	Yes	Yes	Yes
Surviving divorced wife caring for child	Yes	Yes	Yes
Unmarried child or grandchild under 18 or student or disabled	Yes	Yes	Yes
Dependent widower at 60	Yes	No	Yes
Disabled dependent widower at 50	Yes	No	Yes
Dependent parents at 62	Yes	No	Yes
Lump-sum death benefit	Yes	Yes	Yes

(b) Retirement and survivors' benefits

Retirement benefits are paid to a retired worker and to his dependents while he is alive and after his death. The retired worker is entitled to his own retirement benefits at age 62 or over, payments to his wife at age 62 continuing through her widowhood, payments to a child under eighteen, payments to a child over eighteen who is disabled, payments to a child who is a student to age 22 and payments to a spouse of any age if he or she has in his or her care a child who is entitled to benefits. A divorced wife, aged 62 or older and grandchildren may also be eligible for benefits.

Early retirement benefits are 80% of full benefits if retirement is at age 62, 86 2/3% at age 63 and 96 1/3% at age 64.

For each year which a claimant postpones retirement after reaching age 65, his or her benefits will be increased by 1%. The maximum increase is 7% ; there is no credit after age 72.

Children's Benefits

Ordinarily a child's benefits cease at age eighteen except for a child who is disabled before age 22 and whose disability continues thereafter. Benefits are payable to a child who is a full-time student through the semester or quarter in which he becomes 22.

Widow and Widower

Benefits for widows (and dependent widowers) are 100% of the decedent's full benefit. A widow may receive the full widow's benefit at age 62. An unremarried widow of an individual who died fully insured may receive reduced benefits at age sixty. If a widow elects to receive benefits at age 60, the monthly payment will be roughly 13% less than the full benefit allowed if she started at age 62.

Monthly Payments to Survivors

Payments may be made to a widow or surviving divorced wife aged sixty or over, to a widow, widower or divorced wife of any age if caring for a child, to a child or grandchild under eighteen, to a child of any age if disabled, to a child who is a full-time student and to dependent widowers and parents.

Currently insured status alone will provide only benefits for the child, the mother or father caring for it and the lump-sum death benefit. To obtain the others the worker has to be fully insured at his death. A widower can draw benefits on the death of his worker wife if he was dependent and she died fully and currently insured.

Lump-Sum Death Benefits

Death benefits include a lump-sum payment to the spouse or the person who paid the worker's burial expenses in an amount up to $255. If there is no widow or widower entitled to receive the lump-sum death benefit, it may be paid directly to a funeral home or to the person who paid the funeral home and any excess will be paid to those persons equitably entitled to it by virtue of having paid expenses not incurred through the funeral home.

Disability Benefits

The disability insurance benefit is payable to a disabled worker who is fully insured and has had at least twenty quarters of coverage in the previous forty quarters. The benefit is the same for a disabled person as it is for a retired worker. All of the same death and dependency benefits result. When the disabled worker attains retirement age he then falls under the regular provisions for retired workers.

Disability means the inability to engage in any substantial gainful activity by reason of any medically determinable physical or mental impairment which has lasted or can be expected to last for a continuous period of not less than twelve months. Blindness, as such, is not a disability for purposes of benefits for disabled workers or children. It may be, however, the impairment to which the statute refers under the general definition of disability.

Amount of Benefits (as of June, 1976)

Average Yearly Earnings	$923 or less	$3000	$4200	$5400	$6600	$7800	$9000	$10,800	$13,200	$14,100
Retired Worker & Disabled Worker	$107.90	$223.20	$272.20	$319.80	$368.10	$422.40	$452.20	$490.60	$539.10	$556.30
Wife (65 or Older)	54.00	111.60	136.10	159.90	184.10	211.20	226.10	245.30	269.60	278.20
Retired Worker at 62	86.40	178.60	217.80	255.90	294.50	338.00	361.80	392.50	431.30	445.10
Wife at 62 (no child)	40.50	83.70	102.10	120.00	138.10	158.40	169.60	184.00	202.20	208.70
Widow or Widower at age 65	107.90	223.20	272.20	319.80	368.10	422.40	452.20	490.60	539.10	556.30
Widow or Widower at age 62	89.50	185.10	225.70	265.20	305.20	350.20	374.90	406.80	447.00	461.20
Widow or Widower at age 60	77.20	159.60	194.70	228.70	263.20	302.10	323.40	350.80	385.50	397.80
Disabled widow or widower at age 50	54.10	111.70	136.20	160.00	184.10	211.30	226.20	245.40	269.60	278.20
Wife under 65 & one child	54.00	118.00	200.80	277.20	298.20	316.80	339.40	368.00	404.20	417.20
Widowed mother & one child	162.00	334.80	408.40	479.80	552.20	633.60	678.40	736.00	808.80	834.60
Widowed mother & two children	162.00	341.40	473.10	597.00	666.30	739.20	791.70	858.60	943.50	973.50
One child of retired or disabled worker	54.00	111.60	136.10	159.90	184.10	211.20	226.10	245.30	269.60	278.20
One surviving child	107.90	167.40	204.20	239.90	276.10	316.80	339.20	368.00	404.40	417.30
Family Maximum	161.90	341.20	473.00	597.00	666.30	739.10	791.50	858.50	943.30	973.40

Generally, average earnings are figured over the period from 1951 until the worker reaches retirement age, becomes disabled or dies. Up to five years of low earnings, or no earnings, can be excluded. Maximum creditable earnings for Social Security are: $3600 for 1951-54; $4800 (1959-65); $6600 (1966-67); $7800 (1968-71); $9000 (1972); $10,800 (1973); $13,200 (1974); $14,100 (1975); $15,300 (1976); and $16,500 (1977). Average earnings generally cannot reach the amounts listed in the columns from $9000 on until future years. The maximum benefit for a man retiring at age 65 in January, 1976 was $364, based on average earnings in the vicinity of $7050. The 6.4% increase effective in June, 1976, payable in July, increased it to $387.30.

(c) Medical benefits

Medical benefits under Social Security arise from two separate programs. Basic Medicare is compulsory and produces hospital benefits. Supplementary Medicare is voluntary and its benefits are primarily doctors' services.

While both are often referred to as "Medicare," it is important to separate "Basic" Medicare from "Supplementary" Medicare. Basic Medicare is financed through a special tax; it is compulsory and the tax is added to the regular Social Security tax. It provides insurance against the costs of hospital care for persons 65 or over.

Supplementary Medicare is a voluntary program available to persons 65 or over on the payment of a monthly fee. This program provides reimbursement for doctors' bills and some other services not included under Basic Medicare.

Basic Medicare

Benefits are provided to persons who are 65 or over and entitled to monthly Social Security benefits. This does not mean receiving benefits; a person is eligible for Medicare even though his benefits are not paid because he earns too much money after qualification. A person aged 62-64 who is receiving monthly benefits is not yet eligible for medical benefits.

Basic Medicare provides payments for (1) inpatient hospital care, (2) post-hospital extended care, (3) post-hospital home health services and (4) hospital and outpatient diagnostic studies. Payment is normally made to the hospital or other provider of the covered service.

Hospital service benefits cover up to ninety days in the hospital for any "spell of illness" which roughly means the period beginning on the day that a person is first an inpatient and ending at the close of the first period of sixty consecutive days thereafter during which he was not an inpatient. Thus, a person could be in and out of the hospital during a single spell of illness. Two months out terminates the "spell." The services covered are bed and board, nursing (but not private-duty nurses) and other customary patient care, drugs, supplies, equipment, and diagnostic or therapeutic items ordinarily furnished to patients, including blood in excess of the first three pints.

Each person is entitled to a "lifetime reserve" of sixty additional days to be used when he exhausts the ninety days of a spell of illness.

The beneficiary must pay the first $124 during any spell of illness and one-fourth of the cost (or $31) per day for each day over sixty days up to the maximum of ninety days, and $62 a day for each day of the lifetime reserve. In subsequent years these figures will be adjusted for further increases in operating costs.

Supplementary Medicare

An entirely voluntary separate program of medical insurance is provided for people 65 or over who elect to join. A person who has reached 65 or who is entitled to hospital benefits under Basic Medicare may enroll.

Supplementary Medicare is in addition to Basic Medicare. It costs an extra premium and it produces extra benefits. It covers the cost of doctors' bills, additional home health agency visits and other health services not provided by Basic Medicare.

If a person is eligible for Basic Medicare (hospitalization) he or she is automatically enrolled for medical insurance (Supplementary Medicare) unless the enrollment is rejected by the beneficiary.

Medicare Premiums

Premiums are adjusted each year as necessary to maintain the program. Premiums are supposed to carry half of the cost of the program.

The premium for 1976-77 is $7.20 per month. It will rise to $7.70 in July of 1977.

If a person is receiving monthly benefits, the medical insurance premium will be deducted from his benefit. If a voluntary insured is not receiving benefits, he will pay his premiums directly to the Government.

In the supplementary plan, there is also a deductible and a co-insurance feature. The first $60 of the allowed doctors' bills and other services in each year must be borne by the insured and the insured also pays 20% of all additional costs. Thus the insurance pays 80% of the "reasonable charges" over $60.

Medicare for Disabled

Disabled persons are entitled to Medicare benefits if they have been entitled to disability benefits for two years. Fully or currently insured workers and their dependents who have chronic renal disease are disabled for Medicare purposes.

Farm Income Tax Manual, §§ 1018-1021.

chapter 8

TAX ASPECTS OF FARM BUSINESS ORGANIZATION

Section

§ 8.1 BASIC FORMS OF ORGANIZATION

There are three basic forms of business organization: (1) the sole proprietorship, (2) the partnership, and (3) the corporation. A fourth form is a tax hybrid—a corporation taxed like a partnership. Each form of organization has inherent tax advantages and disadvantages.

A farm operator should choose the form of business organization which best fits his operational and estate-planning objectives. Tax advantages may be a motivating factor in selecting a particular type of organization.

The sole proprietorship is the simplest form of organization; the farmer runs his business and reports his taxes as an individual. Management and tax decisions are his alone; they affect no one else.

Partnership and corporate organization add additional parties and create another tax unit separate from the individual partners or shareholders. The partnership tax rules establish a separate tax entity, but not a separate taxpayer; a partnership compiles the tax data, but pays no tax. The members of a partnership report for income tax purposes their prorata shares of the tax determinations of the partnership entity.

A corporation is a separate tax entity and a taxpayer in its own right, separate and distinct from its shareholder-owners who ordinarily receive income only when the corporation distributes money or property to them. In the ordinary small farm corporation, shareholder-owners are also employees of the corporation, renters of property to the corporation, and perhaps creditors of the corporation. Even though a shareholder may be an officer, director, lessor, or creditor, these capacities are different from his status as a shareholder. Farmer Jones' ownership of stock in Jones Farming Company, Inc. is no different than his ownership of stock in General Motors.

The final form of business organization is really a mix of a corporation and a partnership. The Subchapter S corporation or "tax-option" corporation is in all respects a formal corporation organized under state corporation law—but for tax purposes the corporate income is taxed to the shareholders and not to the corporation.

This chapter discusses the partnership, the regular corporation, and the Subchapter S corporation. It emphasizes the details of partnership taxation because farm partnerships are far more common than farm corporations.

The business organization should be carefully selected with the particular objectives of the owners in mind and entered into only with tax and legal counsel to ensure maximum results and benefits.

§ 8.2 FARM PARTNERSHIP

A farm partnership is formed when two or more persons join together for the purpose of carrying on the business of farming. Each partner contributes money, property, labor, or skill, and shares in the profits or losses, in accordance with the partners' agreement.

A partnership is established by the intention of the parties to join together to carry on a trade or business and share in the profits or losses. Intention of the parties is determined from their statements, their partnership agreement, their conduct in carrying out its provisions, the testimony of disinterested persons, the relationship of the parties, the abilities and contributions of each, the control that each has over the income and the purpose for which it is used, and any other facts throwing light on their true intent.

A farm partnership agreement is a legal arrangement which should be in writing. The importance of a written partnership agreement cannot be overemphasized. Of course, it confirms the understanding of the parties, but it also serves to determine many matters of income tax importance.

Joint Business Activities

If an owner of a farm simply leases land to a tenant for cash or a share of the crops, the parties do not ordinarily create a partnership. But if the landowner is active in the business, and profits and losses are shared, the arrangement may be a partnership.

A partnership is often formed by joint owners and operators of real estate, such as heirs or devisees who are owners of fractional interests in farm land. However, mere co-ownership of property which is maintained, kept in repair and leased does not establish a partnership among the owners. If the joint owners of a farm lease the farm for cash or on crop shares, no partnership arises; but if the co-owners actively carry on the farming business and divide the profits and losses, they have formed a partnership.

Parties in a joint enterprise may be partners even though one or more have separate farming income or expenses in which all of the partners do not share. The whole operation could be treated as a partnership and the partnership agreement could allocate particular items of income and expense to one or more partners. It is also possible to treat all shared activities as the partnership business and the rest of the operation as the separate businesses of the parties.

For tax purposes, all income and expenses which are shared would be reported by the partnership. If the partners have a joint interest in all or part of the machinery and equipment, the depreciation schedule on the partnership farm return should include this property. On the partner's individual farm schedule he will list any additional income he might have and all items of depreciation and expense which are a cost to him of producing either his own or partnership income.

Farm Income Tax Manual, §§ 901, 903, 904.

§ 8.3 TAXATION OF PARTNERSHIPS

Every partnership must file a partnership return Form 1065 which reports the shares of partners in the ordinary income or loss and certain other items such as charitable contributions and capital gains. Salary and interest paid to partners are considered expenses to the partnership and income to the recipient partner. Distribution of profits and losses may be changed as the partners may agree before the due date of the tax return.

The partnership pays no tax; the partners report their shares of income and loss on their individual returns.

The partnership income tax return allocates income, loss, deductions and other tax items to the partners. Some items are segregated on the partnership return so that each partner may take his share of those items and aggregate them with similar items from nonpartnership sources. The partner includes in his own income tax return his share of the partnership income whether income is actually distributed or not.

The partnership return must segregate and the partner report his share of capital gains and losses, Section 1231 gains and losses, charitable contributions, soil and water conservation expenses, self-employment income, and other items. If the partnership agreement allocates various items of income, loss, or deduction to specific partners, these items are also segregated on the partnership return. Too, any item that would be an individual deduction but not a partnership expense must be allocated to the partner or partners concerned.

If the partnership suffers an operating loss, each partner claims his proportionate share and, aggregating it with his own nonpartnership income and loss, determines the amount of his own net operating loss available for carry-back and carry-forward. If various items of income and expense are allocated to particular partners, it is possible for a partner to have a loss even though the partnership has taxable income.

The partnership agreement allocating these items is respected except to the extent that the allocation does not have substantial economic effect.

Conduit Theory

The theory that a partnership is a conduit of tax items to the partners controls the entire relation of partnership to partner. The character of any item of income, gain, loss, credit or deduction is the same to the partner as it is to the partnership, just as if the partner had received his share directly. Gross income of a partner is his share of partnership gross income for all necessary purposes. The partner's share of gross income of the partnership is "gross income from farming" for purposes of filing a declaration of estimated tax. When a farm partnership elects to deduct soil and water conservation expenses, the partner uses his share of partnership gross income to determine the limits on the deduction on his own return.

Computation of partnership income follows the usual rules except for the items which are separately reported. There is no real taxable income of a partnership because the allocation schedule is divided into categories. Of course, the percentage standard deduction, low income allowance, personal exemptions, personal deductions, net operating loss and deduction for capital gains do not appear on the partnership return. Each partner takes his own.

Division Among Partners

The agreement among the partners controls the distribution of profits and losses among them unless the agreement allocates items without a substantial economic effect. No set percentages need be adhered to and the partners may, for valid reasons, decide on or before the due date of the return to divide income or losses in a different manner than was previously agreed upon. Of course, a sham division or one without substantial economic effect would be subject to challenge, but, within those limits, the law makes it clear that the partners decide how to divide profits or losses, and they can do it up to the date the return is due.

If there is no partnership agreement, income and losses are shared according to the partners' interests in the partnership. The law does not clearly state how an interest in the partnership is determined but all the facts and circumstances are to be taken into account. This formulation obviously leaves room for second guessing if there is no partnership agreement, and this uncertainty strongly argues for a clear, written agreement. But if there is no agreement, each separate item is shared in the same ratio.

Limitation on Loss

One other limiting factor should be noted. A partner may not deduct a loss in excess of his tax investment in the partnership at the end of the partnership year in which the loss occurred. This means that when a partner's basis for his partnership interest is zero, he cannot claim any losses unless he actually pays into the partnership. It is not uncommon for this situation to arise in a farm partnership where low-basis, high-value property was contributed. For example, if a cash-basis partner's sole contribution was $10,000 worth of raised livestock, the basis for his partnership share or interest would be $0, the basis to him of the animals contributed. See § 8.5. If the partnership had a loss, the partner would have no deduction.

The disallowed loss can be deducted in any subsequent year in which there is a basis for his partnership interest against which the loss can be charged. A partner's basis is increased by contribution of additional money or property, by his share of earnings left in the business, or by his share of partnership borrowings. Under the 1976 Act, in 1977 and later years, a partner's share of the partnership's borrowings does not include any borrowing on which the partner has no personal liability. This limitation does not apply to real estate partnerships or to partnerships subject to the "at risk" provisions discussed at § 4.1. Since farming is subject to these rules, this partnership provision will not apply to farm partnerships, but the "at risk" rules will. Both rules reach pretty much the same result on limiting losses when there are nonrecourse loans.

Accounting Period and Method

The partnership is a separate tax entity. Therefore it selects its own accounting period or taxable year and accounting method, cash or accrual.

A partnership may not adopt or change to a taxable year different from that of the principal partners except with the consent of the Commissioner. A principal partner is one who has an interest of 5% or more in partnership profits or capital.

If the partnership is on the accrual basis, a partner on the cash basis will report his share of partnership income as computed on the accrual basis. In effect, deductions which have accrued to the partnership are allowed to the cash-basis partner.

Elections by Partnership

There is one important exception to the conduit theory. Elections affecting partnership income are made by the partnership, not by each partner. Methods of accounting, computing depreciation, and handling soil and water conservation expenses, for example, are elected by the partnership. However, the partnership election does not affect a partner's nonpartnership activity. Thus, a farmer could deduct soil and water conservation expenditures on the farm in the partnership and capitalize expenditures on another farm not in the partnership.

The partnership makes the election to deduct expenditures for clearing land, but each partner takes into account separately his distributive share of the partnership's expenditures for clearing land and his distributive share of each item of income, gain, loss or deduction required in computing taxable income derived from farming. If the partner has land of his own upon which he has elected to deduct land-clearing expenditures, he must aggregate his own "taxable income derived from farming" with his share of the partnership's similar income and aggregate his separate expenditures and his share of the partnership expenditures for clearing land.

Partnership Return

Form 1065 is solely an information return which reports each partner's share of income or loss and other items; no tax computation is made. The partnership also files a farm schedule and the capital gains schedules with the Form 1065. Partners report the amounts of their shares of partnership items on their own returns.

A partner reports his share of partnership income and other items in his calendar or fiscal year during which the partnership taxable year ends. For example, if the farm partnership taxable year ends February 28 and the partner's taxable year is the calendar year, his share of the income earned by the partnership from March 1 of last year to February 28 of this year would be reported by the partner on this year's return.

Partnership Organization and Promotion Expenses

The 1976 Act requires that expenses incident to organization or promotion or sale of a partnership be capitalized. The partnership may elect to amortize the organization (but not the sale or promotional) expenses over sixty or more months. If the partnership is liquidated before there is full amortization, a loss may be claimed for the balance.

Farm Income Tax Manual, §§ 902, 910, 916.

§ 8.4 FORMATION AND OPERATION OF A PARTNERSHIP

Neither partner nor partnership realizes gain or loss by contribution of assets to the partnership. The basis of the contributed asset to the partnership is the same as it was to the partner. If a partner deals with the partnership other than as a member, he is treated as though he were an outsider in the transaction.

A partner who contributes property to a partnership does not realize any gain or loss on the transaction. The partnership and the other partners have no gain or loss, either. There are a few circumstances where it is possible that gain to a contributor of property will be recognized, for example, if the partnership assumes liabilities in excess of the basis of the contributed property.

Since no gain or loss is recognized, the partnership takes the contributed asset at the basis that the asset had in the hands of the contributing partner. The basis of the asset is not adjusted even though the market value is higher or lower than the basis in the hands of the contributing partner.

In effect, the contribution of an asset by the partner to the partnership is treated as a tax-free exchange of the asset for an interest in the partnership or an increase of an existing interest.

The partnership must use the contributing partner's adjusted basis as its basis for computing depreciation, depletion and gain or loss. However, the partners may agree to adjust the distributive shares of the partners in order that one will not suffer an unfair disadvantage because of the economic effect of this basis rule.

> **Example:** A and B form a partnership. A contributes an asset which has a basis to him of $1000 and is worth $2000 while B contributes $2000 in cash. Assuming a five-year life, economic depreciation on a $2000 asset would be $400 per year of which B would be entitled to half. But tax depreciation on the $1000 basis can be only $200, of which B would be entitled to half. To correct this, the partnership could elect to allocate the depreciation on the contributed asset to B. Then B would have his economic benefit of $200 in depreciation but A would have none.

Pre-Contribution Gain

If an asset which was contributed by one of the partners is sold by the partnership, part of the gain on the sale may represent gain which accrued prior to forming the partnership. This gain may be charged to the contributing partner if so provided in the partnership agreement.

If the asset in the example above is sold by the A-B partnership for $2000, there will be a taxable gain of $1000. If no agreement is in effect, A and B will each report half of the gain including any ordinary income from recapture of depreciation taken by A. If the partnership agreement provides that the pre-contribution gain is to be charged to A, then A will report $1000 and B will report no gain.

Partner Dealing with Partnership

Although a partner normally deals with the partnership in his capacity as partner, nonetheless a partner may deal with the partnership just as a stranger or outsider does. If the partner engages in a transaction with the partnership other than in his capacity as partner, he is treated for that transaction as if he were not a member of the partnership. Loans of money or property, sales and purchases, and the rendering of services are examples.

However, if a partner owns, directly or indirectly, more than 50% of the partnership, losses on sales or exchanges between the partner and the partnership are disallowed. Also disallowed are losses on sales and exchanges between two partnerships in which the same persons own, directly or indirectly, more than a 50% interest in capital or profits.

If there is more than an 80% interest, capital gains on a sale or exchange may be converted into ordinary income. The rule also applies to sales and exchanges between partnerships in which the same persons own more than an 80% interest. For example, when a partnership owned by A and B sold a citrus grove to a partnership owned by the adult children of A and B, the gain realized by the A-B partnership was ordinary income not capital gain because A and B were regarded as the owners of both partnerships. The ownership of the children was attributed to A and B for tax purposes.

Farm Income Tax Manual, §§ 905, 907.

§ 8.5 NATURE OF PARTNER'S INTEREST

A partnership is an entity separate and distinct from its partners. A partner is regarded as owning a partnership share or interest which has a tax basis to him that is separate and distinct from the basis of the assets owned by the partnership.

The partnership is a separate entity which owns property. Each piece of property has a basis in the hands of the partnership. A partner has a basis for his interest in the partnership quite apart from the basis of particular pieces of partnership property.

When a partner contributes property on formation of a partnership, he obtains a basis for his interest in the partnership equal to the money and the aggregate of the adjusted bases in his hands of the property he contributed. If he bought his partnership interest, his basis would be the cost plus his share of partnership liabilities, if any. The share of partnership liabilities is subject to the limitation stated above in § 8.3.

Basis Adjustments

A partner's basis for his partnership interest will change from time to time and it is determined without reference to his partnership equity or capital account. It starts with his original basis on contribution to a new partnership or purchase of an interest or acquisition of an interest by death or gift plus his share of partnership liabilities. The basis of his interest is increased by subsequent contributions—money and the adjusted basis of contributed property. If the partnership borrows money, a partner is treated as having made an additional contribution because he assumes his share of the liability.

The basis is further increased by his share of partnership earnings, taxable and nontaxable. The basis is decreased by distributions of property or money, by his share of partnership losses, and by certain nondeductible expenditures. The basis cannot fall below zero.

Example: Assume that partner A contributed raised cattle worth $10,000 with a basis to him of $0 plus $5000 cash to the

A-B partnership. A's basis for his partnership interest is $5000. A-B partnership operates for one year and has a net profit of $6000 to be shared equally by A and B. A's basis has increased to $8000 ($5000 plus his $3000 share of the earnings). If A withdraws $2000 of his share, his basis decreases to $6000. If A-B partnership borrows $1000 from the bank, A's basis increases by $500 to $6500.

The importance of keeping track of the basis of a partner's interest becomes clear to farm partners on a closing-out sale. The gain or loss on the partnership operation is the difference between the partner's share of the money received on liquidation and the basis for his interest in the partnership. If there is no record of contributions and no record of earnings that were left in the partnership, the partner cannot show a basis to offset the amount distributed to him. It is not uncommon to have an apparent gain on liquidation of a farm partnership turn into a loss when all the additions to basis are tracked down.

Farm Income Tax Manual, § 911.

§ 8.6 SALES, EXCHANGES, AND DISTRIBUTIONS

Capital gain or loss results from sale or exchange of a partner's interest in the partnership, except that his share of unrealized receivables and substantially appreciated inventory will produce ordinary income or loss.

Usually there is no gain or loss on a distribution of assets in kind by the partnership to a partner. Cash distributions in excess of the partner's basis result in gain. No loss is recognized except on final liquidation.

The sale or exchange of an interest in a partnership produces capital gain or loss measured by the difference between the amount realized and the adjusted basis of the partner's interest in the partnership. However, where any portion of the proceeds results from the sale or exchange of the partner's share of unrealized receivables and substantially appreciated inventory, there will be ordinary income or loss. This portion of the sale or exchange is simply carved out for special treatment.

One of the features of partnership tax law is the emphasis on "unrealized receivables" and "substantially appreciated inventory." Wherever these are involved, they produce ordinary income.

"Unrealized receivables" are rights to payments for goods or services which have not been reported as income. For example, if a cash-basis farm partnership has sold hogs but has not received the money, it will not have reported the income. However, at this point, each partner owns a prorata share of the right to the money, even though the gain on the sale is still unreported. Therefore, if the partner sells his partnership interest, he must report his share of the earned gain as ordinary income.

Also included in "unrealized receivables" and treated as ordinary income on sale of the partnership interest is the partner's share of recapturable depreciation and land expenses on property owned by the partnership.

"Substantially appreciated inventory items" means the whole inventory of the partnership if the aggregate market value of the entire inventory exceeds 120% of the aggregate adjusted basis of the entire inventory and, in addition, the inventory exceeds 10% of the fair market value of all partnership property other than money. Inventory includes all items held for sale and all property not capital assets or Section 1231 items. Thus, it includes business accounts receivable and unrealized receivables.

> **Example:** A farm partner sells his one-third interest in a $60,000 partnership for $20,000. His share includes one-third of $15,000 worth of raised hogs which have a basis of $0. The partnership also owns $45,000 worth of land on which there is no recapturable depreciation or land expenses. The $20,000 he receives includes $5000 of ordinary income (for his share of the hogs). The rest of his gain is a capital gain.

One other tax consequence to the selling partner must be recognized. He was a member of the partnership up to the day he sold out; therefore, he has a distributive share of partnership income to that date which must be reported.

Distribution of Property

In the usual case of the distribution by the partnership to a partner of property in kind, not in cash, no gain or loss is realized. The partner receiving the distribution will realize his gain or loss when he disposes of the property received. If the distribution is in cash, the partner will have a gain only if the distribution exceeds the basis for his partnership interest.

There are limited exceptions to this rule if the partnership makes payments to a retiring partner or a deceased partner's successor (see § 8.8) or distributions to a partner of unrealized receivables or inventory items which have increased substantially in value.

No loss is realized on a distribution of property or money while the partner continues as a member of the partnership. However, on the complete liquidation of a partner's interest, he will have a deductible loss only if he receives cash, unrealized receivables and appreciated inventory. If he receives any other property, no loss is reported until the property is disposed of.

These provisions are designed to keep a partner from reporting gain or loss for tax purposes before the true economic result of his overall partnership investment is determined.

A special basis problem arises when a partnership distributes property to a partner either while the partnership continues or upon liquidation. Generally, no gain or loss is reported when the property is received. The gain or loss is realized when the partner disposes of the property distributed to him. Thus, he must have a basis for it.

A distinction is made between current distributions where the partner continues in the partnership and distributions in liquidation of his entire interest or of the partnership. In the case of a current distribution, the partner takes over the basis that the property had in the hands of the partnership. However, this cannot exceed the basis of his partnership interest reduced by any money distributed to him. In a liquidating distribution the basis of the property received is the same as the partner's basis for his partnership interest reduced by any money he received on the liquidation.

Farm Income Tax Manual, §§ 906, 908, 912.

§ 8.7 PURCHASE OF A PARTNERSHIP SHARE

The distinction between the basis of a partner's interest and the basis of partnership property can work to the disadvantage of a partner who buys his interest or to a deceased partner's successor. A special election is available to give him a realistic basis for his share of partnership assets.

The basis rules discussed in prior sections can create a result disadvantageous to a partner who bought his interest or to the estate or successor of a deceased partner. For example, assume that the A-B partnership has as its only assets ten raised cows with a basis of $0 which are worth $2000. Partner A sells his partnership interest to C for $1000. If the partnership sold the cows for $2000, C's share of the partnership income would be $1000. But C paid A the value of the half-interest in the cows when he bought the partnership interest.

A similar result follows on the death of a partner if the estate continues in the partnership. If the taxpayer died before 1977, the estate's basis for the partnership interest is the fair market value of the decedent's interest at death which, in this illustration, would also be $1000 but the death of the partner does not give a stepped-up basis to the assets in the partnership. If the partner died after January 1, 1977, the estate's basis is the decedent's basis. This basis is adjusted for increases in value before December 31, 1976, and also adjusted for death taxes on increases in value after January 1, 1977. This basis may also be more than the estate's share of the partnership's basis for partnership property. See § 5.8.

A special election may be made for the benefit of purchasing partner C or the estate. If the election is made, the new partner obtains a "special basis" equal to the excess of the basis for his partnership interest over his share of the basis of partnership property at the time of entry into the partnership.

Effect of Basis Election

In the example above, if the cows were sold for $2000, the partnership would have a $2000 gain because it has a basis of $0 for the cows. If the partnership had made the basis election for the benefit of new partner C, there would be a $1000 basis to apply against C's share of the income from the cows.

If the animals were dairy cows subject to depreciation, the partnership would have no basis for depreciation. But C has paid cash for his share of the cows. If the election is made, C will take depreciation on his basis for half of the cows even though B gets none.

The illustration is a very simple one. In actuality, the special basis will have to be allocated for C or the estate to each of the partnership properties.

An additional, but more limited, election is available to a new partner even if the partnership does not make the special basis election. If purchasing partner C (or the estate or successor in interest of a deceased partner) receives a distribution of property within two years of the date the partnership interest was acquired, the new partner may elect to have an increased basis for the property distributed.

This limited election does not give the new partner any additional basis for depreciation on property held in the partnership or on sale of any property by the partnership.

Farm Income Tax Manual, §§ 913, 915.

§ 8.8 DEATH OR RETIREMENT OF A PARTNER

Tax results to the estate or successor in interest will depend upon whether the partnership continues or terminates, whether money or property is received, and whether installment payments represent the purchase price of a partner's interest or a share of future income. Payments to a retired partner follow the same pattern.

Upon the death of a partner several things can happen:

1. The interest of the partner may be liquidated by payments from the partnership according to a partnership agreement.
2. The partnership may be wound up under the provisions of local partnership law.
3. The surviving partners may purchase the decedent's interest at the date of death under a "buy-out" agreement.
4. The estate or a successor in interest of the decedent may continue as a partner.

Liquidation by Partnership

If the partner's successor or beneficiary receives liquidating payments from the partnership, a distinction must be made between payments in exchange for the partner's interest in partnership property and other payments. They are handled differently.

Payments made in liquidation of the deceased partner's interest in partnership property will result in capital gain or loss, if any, except for the portion of the payments that are attributable to the decedent's prorata share of unrealized receivables and substantially appreciated inventory. The portion attributable to the receivables will result in ordinary income. Although the portion attributable to the inventory produces ordinary income or loss, under the basis rules discussed at § 8.7 the deceased partner's successor may obtain a new date-of-death basis for the inventory which may eliminate some of the gain. No part of the payments can be attributed to the partner's share of good will and treated as payment for his capital interest unless the partnership agreement provides for good will.

Payments in excess of the value of the deceased partner's interest in partnership property are treated as a continuing share of partnership income, taxable to the decedent's successor in interest as ordinary income and deductible by the partnership in computing its income. This is true whether the amounts are based upon a share of partnership income or are fixed in amount.

For practical purposes, the amount paid in exchange for the decedent partner's interest in partnership property is his prorata share of the value of partnership assets at the date of death. Any additional payments are "other payments."

Winding Up

Winding up is a form of liquidation in which all partnership property or money is distributed to the partners and the partnership is terminated. Such a liquidation of a deceased partner's interest is treated as a distribution to his estate. The other partners also receive a liquidating distribution. See § 8.6.

Sale to Surviving Partners

If the partnership interest is sold to the surviving partners, the transaction is treated the same as any sale. See § 8.6. The estate of the deceased partner will, in cases where death occurred after January 1, 1977, have a basis equal to the decedent's adjusted basis, increased by appreciation in value before January 1, 1977, and by death taxes on increases in value since then. If a loss is involved, no adjustments are to be made. If death occurred before 1977, the basis for the partnership interest will be the value at death. Ordinarily, the purchase price will be the fair market value of the interest at death. When death occurred before 1977, there will be no gain or loss on the sale, except that the decedent's share of unrealized receivables of the partnership will be treated as income in respect of a decedent and will produce ordinary income. If death occurred after 1976, gain or loss would be measured by use of the basis established by these rules. See § 8.7.

Continuation as a Partner

The decedent's estate or successor in interest may continue as a partner. If so, the new partner is treated as if the partnership interest was purchased at the basis determined under the rules just outlined. It may be very important for the partnership to make the basis election discussed in § 8.7 for the benefit of the estate or other successor in interest.

Termination of Partnership

Death does not terminate a partnership nor the taxable year of the partnership or the decedent partner. The partnership will continue with the decedent's estate or successor in interest being treated as a partner for tax purposes even if there were only two partners.

Termination occurs only if the partnership is discontinued and no part of the business is carried on by any of the partners in a partnership or if 50% or more of the partnership is sold or exchanged. Transfer by gift, bequest or inheritance is not a sale or exchange. To terminate a partnership on death it is advisable that a provision in the partnership agreement so provide and that the business clearly cease operation as a partnership. Upon termination, the properties of the partnership are treated as distributed in liquidation.

When the estate is entitled to installment or winding up payments, the estate will continue as a partner until the entire interest is paid off, even in a two-man partnership. Thus the decedent's successor is part of the partnership for tax purposes until the final pay-out.

Partner's Income Share in Year of Death

If the deceased partner's interest is sold as of the day of death, the taxable year of the partnership closes with respect to the decedent and the decedent's share of partnership income to the date of death is reported on the decedent's final income tax return.

Otherwise, partnership income to the date of death will not be reported on the decedent's final return, but the entire year's income share will be reported by the estate (or successor in interest) as if it had been a partner for the full year.

Retirement of Partner

The tax aspects described above in terms of the death of a partner apply generally to payments to a retiring partner except, of course, the retired partner is the taxpayer. Payments to a retired partner for his share of unrealized receivables and appreciated inventory always produce ordinary income.

§ 8.9 FARM CORPORATION—REGULAR

A corporation is a separate legal and tax entity distinct from its shareholder owners and is taxed at corporate rates which are different than individual tax rates.

Shareholders may contribute property to a newly formed corporation in exchange for stock without recognition of gain or loss.

Distributions of corporate income to shareholders are treated as ordinary dividend income taxable to the shareholders even though the income was taxed to the corporation.

Sale of stock of a corporation or the liquidation of a corporation ordinarily produces capital gain or loss to the shareholders.

The continued interest in the small farm corporation warrants some comparison of its tax aspects with those of the partnership (§ 8.3) and the tax option corporation, the so-called Subchapter S corporation. See § 8.10.

Farmers may find, as have other businessmen, that the corporation is a satisfactory vehicle for doing business. The usual farm corporation has only a few stockholders, often members of one or two families, and most of its initial capital comes from the family stockholders. The corporation has some favorable non-income tax aspects that appeal to farmers. It limits liability of the shareholder owners for business debts or losses. Since the corporation converts the farm business into easily transferable shares, it aids in estate planning, particularly as a tool for dividing the business among family members. Too, a corporation does not die, thus it ameliorates some of the problems on the death of the farm owner and helps to pass a going business to the next generation.

Separate Tax Entity

A corporation is a separate tax entity. A "regular corporation," as distinguished from a Subchapter S corporation, (§ 8.10), is a taxpayer in its own right.

A corporation, as a separate tax entity, is taxed at the rate of 20% on the first $25,000 of corporate income, 22% on the next $25,000, and at a 48% rate on all corporate income in excess of $50,000. In the computation of corporate income, the corporation deducts salaries, wages, rents and interest even though paid to persons who are shareholders; such payments are income to the recipients.

If scheduled rate changes are allowed to become effective, the corporate rate in 1978 and later years will be 22% on the first $25,000 of income and 48% on the excess.

One of the important aspects of the corporate tax structure is the possibility of double taxation of corporate income to the corporation and to the shareholders. The corporation pays tax on its income as a separate taxpayer and when it distributes dividends from after-tax income to the shareholders, dividends are taxed to the shareholders as ordinary income. Dividends are not deductible by the corporation in computing its income. Thus, corporate income is subject to tax at the corporate level and again at the shareholder level if dividends are distributed. Each taxpayer has a dividend exclusion which eliminates tax on $100 of dividends from all sources.

The disadvantage of double taxation can often be minimized by paying salaries to shareholder-employees, or rents to shareholders who lease property to the corporation, or interest to shareholders who loan money to the corporation. To this extent, the income is taxed only once. The items are deductible expenses to the corporation and income to the employee, lessor or creditor.

One of the problems of small farm corporations is that of hidden or indirect dividends. Any transfer of value from a corporation to a shareholder is a taxable dividend. Thus, if a shareholder obtains meat or milk, or lives in a corporate owned residence, or uses the company car for personal purposes, the fair market value of the item or service may be regarded as a dividend. If the shareholder is also an employee, the distributions may be extra salary. In some cases, though, it is possible that meals and lodging are provided for the convenience of the corporate employer and may be excluded from income by the employee. See § 4.3.

The corporation, as a separate taxpayer, reports its own losses and its own capital gains. An operating loss of a corporation cannot be used by the shareholders to offset their other income because a corporation cannot distribute a loss. Corporate capital gains cannot be passed to the shareholders; capital gains are taxed to the corporation at a 30% rate. Further, if a corporation distributes capital gain income to its shareholders, the capital gain character is lost and the distribution is an ordinary income dividend to the shareholders.

Thus, the main disadvantages of the corporate tax structure to the farmer contemplating incorporation are: (1) double taxation of corporate income distributed as dividends; (2) less favorable treatment of long-term capital gains; and (3) "locked-in" operating losses of the corporation. Note that these are the three aspects that have been changed in the Subchapter S corporation.

Tax Advantages

If an individual's tax rate exceeds 22% there may be an advantage to doing business as a corporation which can pay the tax on the first $50,000 of corporate income (after salaries, rents, interest, etc.) and reinvest it in the business at a lower tax cost. A sole proprietor or a partner would have that $50,000 added to his income at the regular individual rate brackets.

For example, suppose a married sole proprietor had taxable income of $100,000 of which $50,000 was withdrawn from the business. The top brackets on the amount left in the business run from 50% to 60%. If the business was operated as a corporation which paid him a salary of $50,000, the first $25,000 of the other $50,000 would be taxed at 20% and the second $25,000 at 22%. If the owner is single, there would be even greater tax savings because his own personal rates rise more quickly. Of course, if the shareholder wants to withdraw the reinvested earnings from his corporation, he will receive ordinary dividend income which, of course, defeats his plans for cheaper after-tax dollars. As noted below, he does have a chance to realize the accumulated earnings at capital gains rates if he sells or liquidates.

The tax rates on corporations are scheduled to rise after 1977 when the rate on the first $25,000 is supposed to be 22% and 48% on all amounts over $25,000. If this rate change takes effect as scheduled, these calculations would have to be changed to reflect the new rates.

The corporation as an employer may provide tax-favored fringe benefits for its employees including shareholder-employees. It can establish pension and profit-sharing plans and health and accident plans and buy group term insurance for the employees. The corporation deducts the cost but the benefits are not treated as income to the employee (except for premiums on group insurance coverage in excess of $50,000). A self-employed person may establish a deductible retirement plan but it has fewer tax advantages; medical insurance may be deductible as a medical expense to an individual, but life insurance premiums are not deductible.

The 1976 Tax Reform Act permits an employer to buy prepaid legal services for an employee. Deductible by the employer, but not income to the employee.

Tax Effect on Organization

The organizers of a corporation may transfer farm assets to the new corporation in exchange for stock on a "tax-free" basis if the persons who transfer the property receive at least 80% of the voting stock and at least 80% of the value of all other stock issued. This is the ordinary way that small corporations are formed.

Such a "tax-free" exchange of property for stock avoids any gain or loss on incorporation either to the shareholders or to the corporation. Because there is no gain or loss recognized, the corporation takes as its basis for the transferred property, the basis of the property in the hands of the contributing shareholder. In practical effect, the corporation takes over the depreciation schedule and the tax basis on all items received from the organizers. For example, if a farmer contributed $20,000 worth of raised livestock or other farm products, which had a zero tax basis, he would recognize no gain on the transfer and the corporation would also have a zero basis for the livestock or products. If the corporation sold the items for $20,000, the corporation would report $20,000 of income but the contributing shareholder would have no income unless a dividend was distributed.

The shareholder's stock which he receives in exchange for the contributed farm products will have the same basis as the basis of the property contributed. In the illustration above, the basis of the stock will be zero because the basis of the farm products was zero.

The gain which was not recognized on incorporation will be picked up when the stock is sold. For example, if the incorporator sold his stock for $20,000, he would then recognize his gain of $20,000.

Operation of the Corporation

The farm corporation operates in much the same fashion as a nonincorporated farmer except that the farmer is an employee of the corporation instead of self-employed. Corporate income is computed in about the same way as an individual's income. The corporation is a "farmer" for federal income tax purposes. Therefore, the special rules for farmers are applicable except for a few provisions which relate only to individuals or to certain farm corporations which must use accrual accounting. There are a few items that are handled differently by corporations and individuals, but generally they are not important in deciding whether to incorporate. For example, a corporation is limited to $2000 of extra first-year depreciation. If the operation was a sole proprietorship and a joint return was filed, the owner would be entitled to a maximum of $4000 extra first-year depreciation.

Termination of the Corporation

Although a corporation may be formed without tax, termination is ordinarily taxable. The usual rule on liquidation or termination of a corporation treats the shareholder as if he sold his stock to the corporation for the fair market value of the property distributed to him on liquidation. Some farmers have run into the trap of paying tax on the appreciation in value of the contributed property when they became disenchanted with their corporations and liquidated.

Suppose that a farmer contributed to his corporation a breeding herd that had a value of $50,000 and a basis of zero and farmland that had a value of $100,000 and a basis of $60,000. No gain was recognized on incorporation. Four years later he liquidates the corporation and receives back the breeding herd which has increased in value to $75,000 and the land which has increased in value to $135,000. The value of the property received on liquidation is $210,000 and the basis of his stock is $60,000. His gain on the liquidation is $150,000 on which he will have to pay capital gains tax. He would have paid the same tax if he sold his stock to someone else, but it seems hard to pay the tax to get your own property back.

The moral is that one does not incorporate unless he plans to continue the corporation indefinitely and he does not do so without advice on the tax rules of formation, operation, distributions to shareholders, and liquidation. Ordinarily a farm corporation is not a complicated or troublesome tax creature; it is important to see that it stays that way.

The tax rules apply to corporations large and small, and a claim of ignorance of the tax effects does not work very well either for big General Motors or little Jones Farming Company, Inc.

Most of the agricultural universities and county extension offices have pamphlets on farm corporations. They can be very helpful to farmers thinking about incorporation.

§ 8.10 FARM CORPORATION—SUBCHAPTER S

A farm corporation with one class of stock and ten or fewer stockholders may elect to have corporate income taxed to the stockholders and not to the corporation.

A farm corporation with one class of stock and not more than ten stockholders (fifteen stockholders after 1976 when the corporation has been a Subchapter S corporation for five years) may avoid the "double tax" on corporate profits by electing to be taxed, not really "like a partnership," but in a manner somewhat similar to the method of taxation of a partnership and its partners. It has been estimated that about 10% of the country's two million corporations have made such an election.

While the election is in effect, the corporation pays no income tax on corporate earnings, but each stockholder reports for his taxable year in which the corporation's taxable year ends the income of the corporation actually distributed to him and his prorata share of the corporation's undistributed income.

Corporate long-term capital gains "pass through" to each shareholder who reports his share of the gains on his own return. Each stockholder also reports his share of a corporate loss which can become a net operating loss to him just as it would be in a sole proprietorship or a partnership. See § 4.18.

The stockholder's share of the income from an electing corporation is not derived from a trade or business carried on by the stockholder and, therefore, is not "net earnings from self-employment" subject to Social Security taxes. It is not "gross income from farming" for purposes of declaration of tax. Nor does the share of corporate income reduce Social Security benefits after retirement.

The election often serves as a method of spreading income among the family, including minor children, but the technique will be challenged if it results in transferring payment for a farmer's labor and management to others, or if the transfer of stock is not full and complete.

A corporation which elects the special tax treatment under these provisions does not cease to be a corporation; in all other respects, it is subject to corporate tax rules. For example: the corporation may have profit-sharing or retirement plans although not as favorable plans as a regular corporation; or a medical plan for the benefit of its employees, including shareholder-employees; a shareholder-employee may receive meals and lodging without tax if it is for the convenience of the corporate employer (see § 4.3); and other fringe benefits are available to employees even though they are also shareholders.

Eligible Corporations

To qualify, a corporation must be a domestic corporation with no more than ten stockholders. After 1976, the corporation may have as many as fifteen stockholders if its election has been in effect for five consecutive years. During the five years, it may have from ten to fifteen stockholders if all of those over ten acquired their stock by inheritance. Before 1977, all stockholders had to be individuals or estates. After 1976, voting trusts or "grantor" trusts (where all income is taxed to grantor) may also be stockholders. Any other trust may not hold stock in an electing corporation, but a testamentary trust may own the stock for sixty days without disqualifying the corporation. Stock may be held for a minor by a custodian under the Uniform Gifts to Minors Act or the Gifts of Securities to Minors Act. A life tenant of stock is an eligible stockholder.

A corporation is not eligible to make the election if more than 80% of its gross receipts comes from sources outside of the United States or if more than 20% of its gross receipts comes from royalties, rents, dividends, interest, annuities, and sales and exchanges of stock or securities.

The typical eligible corporation, or "small business corporation," is the ten (or fifteen) shareholder or smaller outfit with only common stock outstanding, engaged in farming, manufacturing, merchandising or the like.

Undistributed Taxed Income

If a stockholder has reported income which has not been distributed to him, he increases the basis of his stock by the income reported. When the tax-paid corporate earnings are actually distributed, the stockholder pays no further tax on the income but reduces the basis of his stock by the amount of the distribution.

The basis of the stock is also reduced by deductible losses. Losses in excess of the basis of the stock reduce the basis of any loans made to the corporation by the stockholder.

Problems

The hybridization of corporation and partnership creates special problems not shared by either of the parents. Of special importance are those dealing with the nature of the income of the corporation, the revocation or termination of the election, and the death of the shareholder.

A corporation may elect the special method of taxation only if 80% of its income arises from operations and not from investment. Farm corporations are particularly concerned with the meaning of "rents." If 20% of the corporate income comes from rents and other investment sources, the election cannot be made, or if made, it will terminate. Rents under crop share or livestock share farm leases are investment income within the 20% rule unless the corporation "materially participates" in the farm operation. Thus, the self-employment Social Security concepts define the income of farm landlord corporations. See § 7.5.

Revocation and Termination of Election

A voluntary revocation of the election can be made by the corporation and its stockholders. Before 1977, an election was automatically terminated if a new stockholder did not consent to continuation of the election within thirty days. After 1976, the election continues unless, within sixty days after becoming a shareholder, the new shareholder affirmatively refuses to consent to the election. Thus, an electing corporation may be at the mercy of a minority stockholder. He can force the termination of the election by transfer of his stock to another who refuses to consent to continuation of the election. An agreement among shareholders prohibiting transfer of any shares unless the transferee consents to continuation may be advisable. Such an agreement may also bind the estate or heirs of a shareholder.

An election terminates if the corporation ceases to meet the definition of a small business corporation, for example, if a second class of stock is issued, or the stockholders exceed ten (or fifteen, where fifteen applies) or the wrong kind of trust, or a corporation or a partnership becomes a stockholder. The election also terminates in any year that the corporation derives more than 20% of its gross receipts from rents, royalties, dividends, interest, annuities and sales and exchanges of stock or securities, or more than 80% of its gross receipts from sources outside the United States.

A new business has a limited protection from the rule that the election ceases if 20% of the income is from passive investment sources. The election is not lost in the first two years even though 20% of the income arises from the forbidden sources, if it is less than $3000.

Inadvertent termination of the election can occur, for example, by a transfer of stock to an inter vivos trust, by a transfer of stock from an executor to a testamentary trust (which retains the stock for more than sixty days), by a will transfer of stock to heirs or legatees which increases the number of stockholders to more than ten (fifteen where applicable), by a shift in income to the forbidden types, or by discovering that rents from farmland do not involve "material participation." The danger in an inadvertent termination of an election is the possibility that it is discovered some years later and the Government claims the regular corporate income tax on the corporate earnings. Properly drafted shareholder agreements will help to prevent this result.

A new election cannot be made for five years after termination or revocation unless the permission of the Commissioner is obtained. Continuation of the delusion of a valid election has no effect.

Death of Shareholder

On the death of a shareholder, no part of the undistributed taxable income of the corporation of the year of death is included in his final income tax return. If the deceased shareholder's successor in interest (estate, heirs, beneficiaries or purchasers of the stock) does not negate the election, the prorata share of the undistributed taxable income of the corporation in the year of death will be taxed to the person who holds title to the stock at the close of the corporation's taxable year. However, a distribution to the new stockholder out of undistributed income of prior years (which was previously taxed to the deceased stockholder) will be taxable since the right to withdraw prior taxed income is personal and does not pass to the new shareholder.

The distributive loss of a deceased shareholder for the portion of the year up to his death is allocated to him and is included in his final income tax return. The successor owner of the stock, for example, the estate, is entitled to deduct its share of the loss for the portion of the year during which the successor actually owned the stock.

Farm Income Tax Manual, § 917.

chapter 9

MINIMIZING TAXATION OF THE FARM ESTATE

Section

§ 9.1 TAXES IN ESTATE PLANNING

Estate tax planning is the process of arranging for the transfer of property to provide the minimum tax burdens during life and at death consistent with the property owner's nontax objectives. Tax planning requires attention to income taxes, gift taxes, Federal estate taxes, and state death taxes.

There are many definitions of estate planning but all of them recognize the basic fact that a person plans the acquisition, enjoyment, and distribution of property for the benefit of his family. Everyone has an "estate plan" whether he knows it or not. If he does not have an organized plan, he has bits and pieces of a plan—the purchase of insurance, the taking of title in joint tenancy, the sale of property on an installment basis, and many other common acts are estate planing decisions. Even the failure to make a will is a form of estate plan because the state law will distribute property on death. Some planning is needed by everyone.

Tax Planning

One of the major factors in estate planning is the impact of taxes. Death taxes can deplete a farmer's estate and reduce the amount of property which can be left to his wife, children, and other heirs. The Federal Government levies a tax on transfers during life and at death and all states except Nevada levy an additional death tax.

Income taxes are also of concern in tax planning. Income tax savings during life create additional capital to be used or passed on. Too, the way a will or trust is drawn will determine how much the beneficiaries after death will have to pay in income taxes.

Gift taxes are levied by the Federal Government and a few states on transfers made during life; these are payments in advance of or in lieu of death taxes.

Many people with small estates fear that taxes will take everything while others with larger estates either don't realize the aggregate value of their property or are not fully aware of the tax bite.

Most of this book has been concerned with income taxes. This chapter sketches the new unified federal estate and gift tax in the context of estate planning. One of the important lifetime tax-planning tools, the Self-Employed Retirement Plan is discussed in § 9.4; see also § 4.24.

Information on farm estate planning devices and how they work can be found in pamphlets published by many state universities. They are very helpful sources of ideas. But no source of written information can substitute for a farmer's own lawyer. Estate planning involves a mixture of many areas of law and taxes; it is important that it be done well.

The Tax Reform Act of 1976 revised the Federal estate and gift tax system very extensively. The discussion below describes these new rules for 1977 and subsequent years. Where relevant, the rules in effect before 1977 are noted.

§ 9.2 UNIFIED FEDERAL ESTATE AND GIFT TAX

Before 1977, the Federal estate and gift taxes constituted a dual system. Gifts made during life were taxed under one rate schedule and property passing at death was taxed under another higher rate schedule.

Under the Tax Reform Act of 1976, the two taxes were unified into a single rate schedule. Taxes on gifts during life are treated as advance payments on the total taxes due at death on all transfers during life and at death.

There is a credit against the combined tax of $30,000 in 1977 which will increase to $47,000 in 1981. The credit replaces the

specific exemption of $30,000 for gifts and $60,000 for estates which existed before 1977.

The marital deduction, which allows a farmer to make tax free transfers to a spouse, was increased by the new law.

The new unified gift and estate tax law taxes all transfers (after 1976) under the same rate schedule. Gifts during life are taxed each year and the rate of tax is determined as if each year's gifts were added to the gifts of prior years to move up to higher rates on the graduated rate schedule.

At death, a tentative tax is computed on the aggregate of all the gifts made during life and all the property passing at death. In effect, this represents a tax on all of the transfers made since 1976. However, since taxes were collected on the gifts made during life, these are treated as advance payments which reduce the tentative tax at death.

The so called "unified" credit may be applied against the gift taxes due during life. Whatever is left of the credit will offset the estate tax due at death.

Thus, the new law looks to all of the donor's transfers (during life and at death) treating each transfer on top of the prior transfer in applying the graduated rate schedule and collecting the final portion of the transfer tax at death.

The credit under the new law allows more property to pass free of tax than the gift tax and estate tax exemptions under prior law.

For comparison purposes, the table below shows the amount of the credit allowed and the amount of property that will pass free of tax as a result of the credit. Note that the credit will be phased in from 1977 through 1980 reaching its peak in 1981.

Year	Unified Credit	Amount Passing Free of Tax
1977	$30,000	$120,666
1978	34,000	134,000
1979	38,000	147,333
1980	42,500	161,353
1981 and later years	47,000	175,625

(a) Federal Gift Tax

The Federal gift tax is levied on transfers made during life. Gifts are taxed under the graduated rate schedule as illustrated below and the tax is paid annually unless gifts in the year exceed $25,000. If gifts exceed $25,000, a gift tax return is due for the quarter in which the gift total reaches $25,000.

The more gifts that a person has made in prior years, the higher the tax rate will be on gifts of the current year. To determine the tax using graduated rates, it is necessary to know the amount of gifts made in all years since 1976 when the unified estate and gift tax started. The gifts of the current year are considered to be on top of the pile of all the gifts made since 1976.

To compute the tax on the gifts made in the current year:

(1) A tax is computed (using the table below) on all of the taxable gifts after 1976 to the end of the current year, and

(2) A tax is computed (using the same table) on all of the taxable gifts after 1976 to the beginning of the current year.

(3) The difference between (1) and (2) is the tax on the gifts made during the current year.

(4) This amount is then reduced by any unified credit remaining after use against other gifts after 1976.

Taxable Amount	Tax
Not over $10,000	18 percent of such amount.
Over $10,000 but not over $20,000	$1,800, plus 20 percent of the excess of such amount over $10,000.
Over $20,000 but not over $40,000	$3,800, plus 22 percent of the excess of such amount over $20,000.
Over $40,000 but not over $60,000	$8,200, plus 24 percent of the excess of such amount over $40,000.
Over $60,000 but not over $80,000	$13,000, plus 26 percent of the excess of such amount over $60,000.
Over $80,000 but not over $100,000	$18,200, plus 28 percent of the excess of such amount over $80,000.
Over $100,000 but not over $150,000	$23,800, plus 30 percent of the excess of such amount over $100,000.
Over $150,000 but not over $250,000	$38,800, plus 32 percent of the excess of such amount over $150,000.
Over $250,000 but not over $500,000	$70,800, plus 34 percent of the excess of such amount over $250,000.
Over $500,000 but not over $750,000	$155,800, plus 37 percent of the excess of such amount over $500,000.
Over $750,000 but not over $1,000,000	$248,300, plus 39 percent of the excess of such amount over $750,000.
Over $1,000,000 but not over $1,250,000	$345,800, plus 41 percent of the excess of such amount over $1,000,000.
Over $1,250,000 but not over $1,500,000	$448,300, plus 43 percent of the excess of such amount over $1,250,000.
Over $1,500,000 but not over $2,000,000	$555,800, plus 45 percent of the excess of such amount over $1,500,000.
Over $2,000,000 but not over $2,500,000	$780,800, plus 49 percent of the excess of such amount over $2,000,000.

Over $2,500,000 but not over $3,000,000	$1,025,800, plus 53 percent of the excess of such amount over $2,500,000.
Over $3,000,000 but not over $3,500,000	$1,290,800, plus 57 percent of the excess of such amount over $3,000,000.
Over $3,500,000 but not over $4,000,000	$1,575,800, plus 61 percent of the excess of such amount over $3,500,000.
Over $4,000,000 but not over $4,500,000	$1,880,800, plus 65 percent of the excess of such amount over $4,000,000.
Over $4,500,000 but not over $5,000,000	$2,205,800, plus 69 percent of the excess of such amount over $4,500,000.
Over $5,000,000	$2,550,800, plus 70 percent of the excess of such amount over $5,000,000.

Gifts made between September 8, 1976, and January 1, 1977, when the $30,000 specific exemption was available, reduce the unified credit by an amount equal to 20% of any exemption applied to such gifts. Only $6000 of the unified credit may be used against gifts made between January 1, 1977, and June 30, 1977.

Exclusion

Every person is entitled to make tax-free gifts of $3000 per year to each of as many beneficiaries as he wishes. This is the annual exclusion. When the annual exclusion is coupled with the "split-gift" provisions discussed below, a married farmer can make tax-free gifts of $6000 per year to any number of other persons. Before the 1976 Act, the farmer could use the gift tax marital deduction and the annual exclusion to give his spouse $6000 per year tax free. Beginning in 1977, however, the annual exclusion of $3000 applies to a gift to the spouse, and there is also a marital deduction which permits tax free gifts to a spouse until the gifts in excess of the annual exclusions total $100,000.

Marital Deduction

The marital deduction for gifts to spouses makes the first $100,000 of lifetime gifts to a spouse after 1976 exempt from gift tax. When gifts to a spouse have totaled $100,000, the entire next $100,000 is subject to tax. After the $200,000 level is reached, one-half of each further gift to the spouse is subject to tax and one-half is free of tax.

Thus, after 1976, a taxpayer may give $100,000 to a spouse without incurring a tax. The gifts which count against this $100,000 are only those gifts which exceed the annual exclusion of $3000. However, once such gifts reach the $100,000 level, every dollar of gift thereafter over the $3000 annual exclusion is taxable until the $200,000 level is reached. Then only one-half of each dollar of gift over the annual exclusion is taxable.

Split Gift

If a married person makes a gift to someone other than his or her spouse, the gift may be treated as if half were given by the husband and half by the wife. It does not matter who owned the property. In effect, this doubles the annual exclusions and the unified credit, because each spouse is entitled to a $3000 annual exclusion for each donee and each is entitled to the full amount of the unified credit.

For example, suppose Mr. and Mrs. Farmer gave $78,000 equally to their three sons and agreed to split the gift as if Mr. Farmer had given $39,000 and Mrs. Farmer had given $39,000. Since there are three donees, each donor has three exclusions of $3000 which eliminates $18,000 from gift tax. The remaining $60,000 will be split equally between Mr. and Mrs. Farmer. A tax on $30,000 of gifts by each will be computed, and each may use his or her own unified credit against the gift tax if the credit has not been used against other gifts.

There are some technical rules for the use of the exclusion, the marital deduction, and the split-gift provision but the discussion above illustrates the principles of making tax-free gifts.

(b) The Federal Estate Tax

The Federal estate tax is levied on the value of all property passing at death, less debts owed, expenses of the decedent's estate, the marital deduction, and any charitable deduction. The result is the "taxable estate." It is added to the amount of taxable gifts since 1976, and a tax is computed on the total using the rates in the table above. The amount of tax paid on the gifts is then subtracted from this total, and the amount remaining is the estate tax due.

The amount of the unified credit which has not previously been used to reduce a gift tax liability is applied to reduce the estate tax.

After the tax has been determined, certain credits are allowed which further reduce the tax. There is a small credit for part of any state death taxes paid, which most estates will use. A credit is also allowed for foreign death taxes paid.

Another credit is allowed if property is subject to estate tax more than once in ten years. If a father left property to a son upon which estate tax was paid and the son died soon after, the son's estate would be entitled to a credit for the tax paid on the same property in the father's estate.

Property given away before 1977 may be included in the gross estate and subjected to estate tax. If a gift tax was paid on the lifetime gift, it is credited against the estate tax so that the property is not taxed twice.

State Death Taxes

Each state, except Nevada, levies some form of death tax. The details vary from state to state. Some of the taxes are estate taxes, much like the federal estate tax. Most are inheritance taxes which are levied on the amount of property passing to each beneficiary. The exemptions from the tax and the rates of tax tend to vary according to the family relationship of the beneficiary and the decedent. Many state universities have pamphlets discussing the death taxes of their states.

Gross Estate

The "gross estate" includes not only the property owned at death but also property transferred during life if the transfer was a substitute for a death transfer, for example, gifts over which some interest or control was retained.

Specifically, the gross estate includes:

(1) All property owned by the decedent at death;

(2) Life insurance if the decedent owned the policy or if the proceeds were paid to his estate;

(3) Except for certain joint tenancies between spouses, property held in joint tenancy—less any portion contributed by the surviving joint tenant;

(4) Property given away within three years of death, plus the amount of gift tax paid on it;

(5) Property given away during life if the decedent retained a life interest or control over the income;

(6) Property given away during life if the decedent retained other interests, controls, or powers over the property;

(7) Property over which the decedent had a general power of appointment by which he could have taken the property for himself; and

(8) The value of payments to a survivor under certain annuity contracts and retirement plans.

Only half of property held in joint tenancy or tenancy by the entirety by spouses is included in the gross estate if the creation occurred after 1976 and was treated as a taxable gift. For real estate, that means the spouses must have elected to treat the creation of the joint interest as a taxable gift.

Deductions

From the gross estate, there is deducted:

(1) Debts, funeral expenses, costs of administering the estate, and certain casualty losses, such as fire and theft, that occur during settlement of the estate;

(2) "Charitable deductions"—the amount of money or property left to charitable, religious, or educational organizations; and

(3) The "marital deduction."

The Marital Deduction

The marital deduction provides that the estate may deduct the amount of money or property passing to a surviving spouse up to the greater of (1) $250,000 or (2) 50% of the adjusted gross estate. The adjusted gross estate is the gross estate less the items listed as deductions in subparagraph (1) above—debts, funeral expenses, etc. If more than 50% of the adjusted gross estate passes to the spouse, the deduction is limited to the greater of $250,000 or 50%. For example, if the adjusted gross estate was $400,000 and the wife received it all, the limit on the marital deduction would be $250,000. If the estate was $900,000, the limit would be $450,000. No deduction in excess of the amount passing to the spouse is allowable. The effect of the marital deduction is to pass $250,000 or half of the estate to a spouse free of tax.

Orphan's Exclusion

The 1976 Act added a further deduction called the orphan's exclusion. If the decedent is not survived by a spouse but is survived by a child or children without a living parent, there is a deduction for property passing to minor children. For this purpose, "minor" means under 21 years. The limit on the deduction for each child is $5000 multiplied by the number of years by which 21 exceeds the child's age. Thus, if a child is fourteen, up to $35,000 of property which passes to the child may be deducted.

Retirement Annuities

Pension or profit-sharing benefits under qualified employee plans have long been excluded from the gross estate. Under the new law, qualified retirement plans for the self-employed and qualified individual retirement accounts also pass free of tax. The exclusion does not extend to benefits from any of these if they are paid in a lump sum.

Joint Interests

Before the 1976 Act, joint interests were included in the estate of the person who paid for them even though there may have been a gift tax paid when they were created. Under the 1976 Act, new rules may apply to joint interests or tenancies by the entirety created between spouses after 1976. If the creation of such a joint interest resulted in a taxable gift, then only half of the interest will be included in the decedent's estate. If there is no taxable gift, then the new rule does not apply. For example, establishment of a joint checking account does not result in a gift because no gift is made until the donee tenant withdraws funds from the account. The account would be wholly included in the estate of the person who contributed the cash. Purchase of a stock in joint tenancy does result in a gift, and one-half would be included in the estate of each spouse. Creation of a joint tenancy in real estate is treated as a taxable gift only if the donor makes an election to have it treated as a gift. If this election is not made, then the old rule on joint tenancies between spouses will apply. If the election was made, the new rule will apply. This new rule applies only for joint interests held by spouses. Interests not held between spouses do not qualify for the new treatment.

Valuation of Farm

Perhaps the most significant feature of the 1976 Act is the alternate method of valuing real estate used in farming or other businesses. If real estate passes to a family member who continues the use of the property for farming purposes, the real estate may be valued by reference to its use as a farm rather than at its highest and best use. This valuation technique may not reduce the value of the estate by more than $500,000. Once that limit is reached, there will be no further reduction in the estate tax value.

> **Example:** Real estate has a value of $800,000 as subdivision land. The value as farmland determined under a statutory formula is $400,000. This lower figure may be used as the estate tax value. If the highest value had been $1,200,000 with a farm use value of $400,000, the estate tax value would be $700,000 because this preferential valuation method may not reduce the estate tax value by more than $500,000.

If the family does not continue to own the property and use it in the family business for more than ten years, all of the tax savings from the preferential valuation will be recaptured. If the land is disposed of or the use changes from the tenth to the fifteenth year, then the amount recaptured is reduced. The amount of the recapture of the tax savings from the special valuation is reduced by one-sixtieth for each month the land was held over ten years. After fifteen years, there is no recapture. There is no recapture on the death of the heir.

Generation Skipping

Before the 1976 Act, if an estate was very big, a generation-skipping trust could be used to lower taxes. For example, when a parent died, some of the estate might be left in trust to pay the income for life to the children with a power in the trustee to distribute corpus to them if they needed it. On the children's death, the corpus would be paid to the grandchildren. The children had very close to full ownership of the property because they got the income, and the principal could be used for them if the trustee so desired. The children might even be trustees if the power to consume principal was limited to support or other well defined needs. But the property would not have been included in the children's estates because they did not own it.

Under the 1976 Act, the property will be taxed on the children's death. There is an exception of $250,000 for each child. Thus, taxation may now be postponed one generation to the extent of $250,000 for each child (not grandchild). Other generation-skipping trusts are supposed to be taxed once each generation.

§ 9.3 TAX-SAVING PRINCIPLES

The fundamental principles of tax savings are not very complicated. Most tax oriented estate plans will use some or all of these ideas:

1. **Lifetime gifts save income taxes by spreading income among additional taxpayers;**
2. **Lifetime gifts still save estate taxes but not as much as they used to;**
3. **The estate tax marital deduction passes the greater of $250,000 or half of the estate free of tax to a surviving spouse;**
4. **Successive estates may be used even though tax savings from generation skipping have been reduced.**

Estate planning will be profoundly affected by the 1976 Act. The income tax benefit from income splitting from lifetime gifts (see § 2.3-2) will remain. But the estate tax considerations will differ.

Marital Deduction

Lifetime gifts between spouses are now entirely exempt from tax up to $100,000. When this is combined with the enlarged marital deduction for estate taxes, there may be considerable taxes to be saved by proper use of these deductions along with the unified credit which is being phased in over the next five years. In the graduated rate structure, the aim is to create two piles of property which will yield the lowest tax rate rather than having one big pile taxed when the first spouse dies and then another big pile taxed when the other spouse dies. However, each spouse should have as big a pile as can be absorbed against the unified credit.

This technique was standard estate planning before the 1976 Act, and it will remain so after the 1976 Act, but perhaps with some new wrinkles. The common "two-trust" will is based upon a division of the family property into two taxable piles. The results are phenomenal as shown in the example.

> **Example:** A farmer has an estate of $525,000, a wife, and two children who want to succeed to the farm property and operate it. This value gives effect to the special valuation rules for farms and to deductions for expenses of administration and the like. He wonders whether he should leave all of his property to his wife who will then leave it to his children on her death or whether he should leave part of it to her at his death and part to the children. She will then leave the balance to the children on her death. The tax aspects of each plan are as follows:

	All to Wife Wife to Children	Half to Wife Half to Children Wife to Children	All to Children With Life Estate to Wife
Tax on husband's death if he is first to die:			
Adjusted gross estate	$525,000	$525,000	$525,000
Marital deduction	262,500	262,500	—0—
Taxable estate	$262,500	$262,500	$525,000
Tax on taxable estate	$ 75,050	$ 75,050	$165,050
Less unified credit	47,000*	47,000*	47,000*
Tax	$ 28,050	$ 28,050	$118,050
Tax on wife's death			
Gross estate	$497,950	$262,500	—
Expenses of administration	12,950	7,500	—
Taxable estate	$485,000	$255,000	—
Tax on taxable estate	157,300	71,500	—
Less unified credit	47,000*	47,000*	—
Tax	$110,300	$ 24,500	—0—
Tax on both deaths	$138,350	$ 52,550	$118,050

This tax difference is great. If all of the property is left to the wife, the two estates bear a tax of $138,350, but only $52,550 if one-half is left to the children. By leaving one-half to the children, the savings is $85,800.

---

* This is the amount of the credit when it is fully phased in. In 1977, 1978, 1979, and 1980 smaller credits would apply. The computation and principle remain the same.

Successive Estates

If the farmer with the $525,000 estate wants his wife to have the use of all of it, must he pay $85,800 for the privilege by leaving it all to her and having all of it subjected to tax once again on her death, subject to a credit for taxes on his estate in some cases. This suggests the last tax-saving principle, the use of successive estates.

"Successive estates" is nothing more than a fancy term for the old-fashioned will that left the farm to mother for life, remainder to the children. Mother would have the use of all of the property, but in these days of the marital deduction, the tax cost is too high to leave all the property to mother for life. All the property would be taxed at father's death, and none of it would be taxed at mother's death because she would not own any property at her death. The tax cost on father's death would be $118,050 which is also expensive. Is there a solution?

The ordinary answer is a combination of the marital-deduction principle and the successive-estates principle with the plan to provide mother with the use of all of the property.

Half of the property is left to the spouse outright (or it can be in trust) so that it qualifies for the marital deduction. This half will not be taxed on the husband's death but it will be taxed on the wife's death. The other half is left to the wife for life (ordinarily in trust) with a right to all of the income for her life. This half will be taxed at the husband's death but not at the wife's death. Thus, the two halves are separately taxed for maximum tax savings, and the widow has full ownership of one half and all the income from the other half. If her life interest half is in trust, the trustee can be authorized to pay principal as well as income to her without making that half taxable in her estate. In fact, it is also possible to give her a right to withdraw some of the principal without asking the trustee—and still avoid tax on her death.

What has been described above is much like the so-called "two-trust" will, a common estate-planning device. It is a pretty good compromise to save $85,800 on the transfer of a $525,000 farm operation. As the estate increases, so do the tax savings.

Lifetime Gifts

All of the foregoing assumes that the wife survives the husband. If she dies first, then the estate tax on the husband's death will be $118,050 when the property passes to the children. It is the same as the column "All to children with life estate to wife." This estate tax cost may be cut with the new gift tax marital deduction. The husband can give the wife $100,000 and suffer no gift tax cost. If she dies first, her estate pays no tax, and the husband's estate will have its tax cut to $87,300, or a savings of $30,750 over the tax if the gifts had not been made.

While this saving looks good, it must be compared to the total taxes if the gifts are made and if the husband dies first. When we do this, we find that unless the husband limits the amount passing to his wife at death to something less than the full estate tax marital deduction, the wife's estate will be larger than the husband's. Two uneven piles are taxed at more than two even piles. However, the total taxes may be limited to the same amount that would have been paid if no lifetime gifts had been made. To accomplish this result, the husband limits the amount passing from his estate to the wife to one-half of his estate less one-half of the lifetime gifts. This failure to utilize the estate tax marital deduction in full was offset by full use of the $100,000 gift tax marital deduction. Then when he dies first, the estate tax burden on the two deaths will not increase over what it was without the lifetime gifts. If, however, at death, he gives the wife property equal to the full marital deduction, there will be an increase of total estate taxes because the wife's taxable estate will be larger than his taxable estate.

The question which next arises is whether he should make lifetime gifts of more than the $100,000 gift tax marital deduction. He should not from a tax standpoint. One-half or more of all gifts above $100,000 will be subject to gift taxes, and he will have to use his unified credit to eliminate the gift taxes. Yet, to keep taxes on the wife's estate at the lowest rate, he will also have to reduce the amount passing to his wife at death and thus the estate tax marital deduction will not be fully used.

Note on Joint Tenancy

Because of the prevalence of joint tenancy holdings among farmers, it is well to ask whether all of this works if property is held in joint tenancy or tenancy by the entirety between husband and wife. The answer is no! At least for joint tenancies created before 1977. If the farmer in the illustration above holds the $525,000 worth of property in a pre-1977 joint tenancy with his wife (all of which he paid for), the tax result is the same as the first column "All to wife . . ."

All $525,000 of joint tenancy property will be included in the farmer's estate at his death. Half of the adjusted gross estate will qualify for the marital deduction and the federal estate tax on his death will be the same $28,050. However, on the death of the joint-tenant husband, all of the property passes to the surviving joint-tenant wife. On her death, the tax is the tax shown in the "All to wife . . ." column, $110,300.

Can a family holding property in pre-1977 joint tenancy ever achieve the tax savings of column two? Not unless they get out of the joint tenancy. Taking them out raises some very difficult tax questions. Often, it can be done, perhaps at some current tax cost.

The problems arise from a mix of tax rules which are more detailed than can be fully described here. But some comment may give the flavor of the problem. There are three possible solutions: (1) transfer some or all of the property to the husband; (2) transfer some or all of the property to the wife; or (3) divide the property to eliminate the right of survivorship.

A transfer of the husband's half of the property to the wife will be a gift and may be subject to gift tax. The first $100,000 is tax free; the next $100,000 is fully taxable; thereafter only one-half is taxable. The gift tax if all the property is transferred to her is $28,050. If the chances are good that the wife will survive the husband and she owns all of the property, there will be no tax on his death, but her estate will be taxed at about $110,300.

A transfer of the wife's half of the property to the husband runs into some different rules. If the property held in joint tenancy is land acquired after 1954, there is no gift on a transfer back to the one who paid for it unless there was an election to treat the creation of the joint tenancy as a gift. However, a transfer to the husband of the wife's joint interest in land acquired before 1955 or any personal property will be a gift.

If the jointly owned property is divided or converted to a tenancy in common without the survivorship rights, no income tax gain is recognized on the exchange of one form of half interest for another, but another set of gift rules applies. If the property is land acquired after 1954 and paid for by the husband, the division is a gift by the husband to the wife of half of the property unless the creation of the joint tenancy was treated as a gift. The division of any other property is not a gift because it is an exchange for value.

There is one further gift-tax problem lurking in this situation. Ordinarily, if a husband buys property and takes the title in joint tenancy with his wife, he makes a gift to her of half the value of the property (except for land acquired after 1954). In many cases, no gift tax return was filed and no gift tax was paid; if a tax was due years ago, the liability (plus interest and maybe penalties) is still outstanding because no tax statute of limitations ever runs if no tax return was filed.

If the joint tenancy was created after 1976, only one-half of it will be subject to estate tax on the husband's death unless real property is involved. If real property is involved, all of it will be included in the gross estate unless the husband elects to treat half of it as a gift to his wife. If he does, then only one-half will be included in the husband's estate. In either case, all of it will be taxed again when the wife dies. While there sometimes would be a credit against the wife's estate taxes for estate taxes paid by the husband's estate, this credit may not be a full credit, and the tax cost will be higher than if the wife had not owned all the property at her death.

A number of details of the tax law not mentioned in this short discussion and the facts of a particular case may make the problems of getting out of joint tenancy easier or harder.

But it should be clear why lawyers shudder when farmers who have all their property in joint tenancy want to talk about estate planning. And it should also be clear that if the value of property held in joint tenancy reaches into the estate tax brackets, something must be done soon.

Summary

In its simplest sense, tax planning is the technique of figuring out the cheapest tax cost of achieving the family goals. To return to the illustration of the farm owner who has a wife, two children and $525,000 and who wants to leave his property to his wife and children, what are the choices and what are the tax costs?

Here are five ways to achieve his goals—and the estate tax cost on both deaths:

(1) Leave the whole estate to the wife for life, remainder to the children. Tax cost: $118,050.

(2) Leave all the property to the wife outright and let her leave it to the children. Tax cost: $138,350.

(3) Leave half outright to the wife and half outright to the children. Tax cost: $52,550.

(4) Leave half outright to the wife and half to the wife for life, remainder to the children. Tax cost: $52,550.

(5) Give the wife $100,000 during life, leave her an amount which when added to the lifetime gifts is equal to one-half the total of the estate and the lifetime gifts, balance to the children. Tax cost: $52,550.

A well-organized estate plan will use a combination of tax-planning principles. The skill of the estate planner lies in his ability to use the tools of estate planning to achieve the optimum combination of family desires and tax minimization.

§ 9.4 SELF-EMPLOYED RETIREMENT PLANS

The farmer as a self-employed person may deduct payments made to a retirement plan for his own benefit up to $7500 a year. If he has regular employees, the plan must include all of them. Or

he may deduct payments up to $1500 for an individual retirement account or up to $1750 total for an individual retirement account for himself and an equal one for his spouse. Income earned by the retirement fund is not presently taxed, but is deferred until the retirement income is withdrawn.

A plan for his own retirement may be established by a self-employed person, including a farmer or farm partner. The arrangement may be an annuity, pension, or profit-sharing plan. The plan must meet certain qualifications established by the tax law and it must be in writing. A self-employed plan provides tax-deductible contributions for the benefit of so-called "owner-employees," i.e., sole proprietors, and partners who have a 10% or more interest in capital or profits of a partnership.

An owner-employee may not participate in a plan unless it includes all of his full-time employees with three years of service. Only part-time (under twenty hours a week) or seasoned or temporary workers (less than five months per year) may be excluded. The rule applies if there is a single qualified employee.

If several businesses are owned or controlled by the same proprietor or partners, the employees of all of the businesses must be covered and there may be no discrimination among the employees covered by different plans.

Amount of Contribution

A self-employed individual may contribute to a plan any amount up to 15% of his "earned income" but not in excess of $7500. A minimum deduction of $750 is allowed even if it exceeds 15% of earned income. Contributions may be made only if services have been performed. An inactive partner may have self-employment income but he is not qualified for a retirement plan. "Earned income" is basically the net earnings from self-employment. Different rules apply for years before 1974.

Although a farmer is engaged in several trades or businesses, his aggregate deduction for his own benefit cannot exceed $7500. An employee who is also self-employed may claim a deduction based on his self-employment income even though he also participates in his employer's plan.

Amount Deductible

The full amount of the contribution for the farmer is deductible up to the $7500 limit. In addition, the full amount of the contribution for employees is deductible.

Such a plan is worth considering by every farmer. No income tax will be paid on the earnings of the funds set aside. Payments from the retirement plan will be taxed when received on retirement at which time the farmer should be in lower tax brackets. Plans may be integrated with Social Security under certain conditions.

Terms of the Plan

Certain special requirements must be met, in addition to the inclusion of all full-time employees with over three years of service, to qualify the plan for deductible contributions. The plan may not discriminate in favor of the owner-employee. Contributions for the benefit of the employees must vest immediately; if the employee leaves the employment he is entitled to the contributions made for him.

Benefits to owner-employees may not begin before age 59 1/2 except upon permanent disability or death. Payments must commence no later than age 70 1/2 for an owner-employee, or age 70 1/2 or the year of retirement, whichever is later, for others. The funds set aside may not be used for any purpose until retirement. Penalties are provided for premature distributions or improper use of assets or for improper excess contributions by an owner-employee. Transactions between the fund and the owner-employee are generally forbidden.

Funding

A self-employed retirement plan may be funded by:

(1) Contributions to a trust of which a bank is trustee;

(2) Purchase of life insurance, annuity or endowment contracts from an insurance company, with or without a trustee or custodian;

(3) Purchase of stock of mutual funds or regulated investment companies to be held by a bank as custodian;

(4) Purchase of special government bonds, nontransferable and nonredeemable until age 59 1/2, or upon disability or death, with or without a trustee or custodian; or

(5) Purchase of "face amount" certificates which are nontransferable if owned by a person other than a qualified trust.

Employee Contributions

A plan may provide for employee contributions in addition to the deductible employer contribution. The owner-employee may contribute for himself an additional amount equal to his deductible contribution so long as it is at the same rate as that of his employees. While there is no deduction for the extra amount, the earnings thereon are not taxed until distributed. If there are no employees, no extra contribution can be made by a self-employed person.

Taxability on Distribution

Retirement payments are taxed under the regular annuity rules of the Internal Revenue Code. A special averaging device mitigates some of the tax burden upon a lump-sum distribution to an owner-employee if made after age 59 1/2 or upon death or disability.

Establishment of the Plan

A plan may be established during any taxable year and a deduction taken for the contribution made. Various farm organizations and insurance companies have obtained Internal Revenue Service approval of prototype plans. The farmer who desires to enter into a self-employed retirement plan will have no difficulty in participating in a plan with a minimum amount of effort. These previously approved plans require an agreement by the farmer to abide by a plan and pay his money. The local Internal Revenue Service office has official pamphlets on Self-Employed Retirement Plans.

Individual Retirement Accounts (IRA's)

The 1974 law also added a deduction starting in 1975 for Individual Retirement Accounts for taxpayers not covered by a qualified plan, a government plan, or certain annuities. Both employees and self-employed persons receiving compensation will be able to establish individual retirement savings programs if they are not covered under qualified plans of their employers, government plans, or certain annuities. Contributions under the programs will be deductible up to the lesser of $1500 or 15% of compensation. Beginning in 1977, the taxpayer may establish one account for himself and one for his wife. If so, the combined deduction may run to $1750.

INDEX

References are to sections

References are to sections

References are to sections

E

References are to sections

References are to sections

L

References are to sections

References are to sections

References are to sections

References are to sections

References are to sections

U

W